Environmental Security and Global Stability

Environmental Security and Global Stability

Problems and Responses

Edited by Max G. Manwaring

LEXINGTON BOOKS
Lanham • Boulder • New York • Oxford

LEXINGTON BOOKS

Published in the United States of America
by Lexington Books
A Member of the Rowman & Littlefield Publishing Group
4501 Forbes Boulevard, Suite 200, Lanham, Maryland 20706
www.rowmanlittlefield.com

PO Box 317
Oxford
OX2 9RU, UK

British Library Cataloguing in Publication Information Available

Library of Congress Cataloging-in-Publication Data

Environmental security and global stability : problems and responses / edited by Max G. Manwaring.
p. cm.
Includes index.
ISBN 978-0-7391-0447-7
1. Environmental policy—International cooperation. 2. Environmental protection—International cooperation. 3. Global environmental change—International cooperation. 4. Sustainable development—International cooperation. I. Manwaring, Max G.

GE170 .E586 2002
363.7'0526—dc21

2002009451

Printed in the United States of America

™ The paper used in this publication meets the minimum requirements of American National Standard for Information Sciences—Permanence of Paper for Printed Library Materials, ANSI/NISO Z39.48-1992.

Contents

Foreword

Once the realm only of the scientist and those of the same ilk, who read and review refereed technical and scientific journals, environmental security has grown in importance to the point that lofty generals and admirals responsible for regional security now speak of it often. Once the realm only of a few small nongovernmental organizations (NGOs) determined to make a statement from their puny platforms, environmental security has become a topic of which leaders of countries and transnational corporations now speak in every kind of forum. Once ignored in the formulation of national security policy and strategy, environmental security today is addressed in every serious think piece written on any region of the world.

We have arrived, a little late, but together. Commanders of regional military commands, diplomats, policy wonks, oceanographers, foresters, hydrologists, special interest groups, and congressmen and presidents are all thinking about internal and cross-border instability all around the world. When these people look at the root causes of instability that might fester into conflict and possible state failure, more and more fingers point at natural and man-made causes and effects of environmental degradation. They know that no person and no country is immune from these stressors and their associated threats. They are all thinking about how to prevent that instability, and how to cope with it when it does occur. Finally, they all know that, sooner or later, the United States will be faced with unpleasant and costly decisions regarding what to do and how to do it.

Pick your region—the chapter authors who follow have. Pick your poison—the chapters that follow cover many topics that make up the broad categories of "environmental degradation and stressors." The authors discuss the consequences of aberrant water supply (too much and too little), deforestation, pollution, population migration, disease vectors, political consequences of natural and man-made disasters, and more. The contributors to this volume will stir you up. They clearly link cause (environmental irresponsibility) with effect (instability that can lead to armed force intervention). You will see that security-stability based issues—especially those with roots in the environment—are key issues that policy writers, diplomats, warriors, and national leaders must address for security purposes.

In the past, we citizens of the world characterized the environmental issue in terms of human responsibility and civic morality. Rightly so, but too few paid attention. Well, pay attention now. The focus of this volume is security—your security. The people who write know the subject because they have seen its creeping virulence up close. They are experts, hardened by many diplomatic, policy, and security trials.

Reader, please enjoy this important book. Heed the message Dr. Max Manwaring pulls together for you. This subject will increasingly consume the national security debate.

Vice Admiral Paul G. Gaffney II, U.S. Navy
President, National Defense University
Washington, D.C.

PREFACE

Environmental Security and Global Stability: Rethinking Problem and Response

Max G. Manwaring

This book stems from conversations over a period of more than a year with Mrs. Darci Glass-Royal and Commander Ray Simmons, U.S. Navy (Ret.), at the Foundation for Environmental Security and Stability, and subsequent conversations with General Charles E. Wilhelm, USMC (Ret.), and General Anthony L. Zinni, USMC (Ret.), in Washington, D.C. The driving concept behind those conversations is something to the effect that:

> As the United States moves into the next century, global security and stability will increasingly depend on understanding complex and ever-changing geopolitical environments. In that connection, a key element and root cause of the instability and violence that characterizes the "new world disorder" is environmental degradation. In these terms, public policy formulation must account for the global dimensions of environmental security as a stability issue. Anything else would be irrelevant.[1]

Some of the conversations that Mrs. Glass-Royal and I had with General Zinni, and I with General Wilhelm, are important and instructive. Their extensive experience and wisdom inform the development and direction of this work. As examples:

Question: There appears to be no widely accepted agreement regarding the importance of and definition of the environment as a security issue. Is there a definition you find useful?

General Zinni: I would broadly say that when environmental conditions, either natural or man-made, are destabilizing a region, a country, or have global implications, then there are major security implications. Those problems could affect stability, could result in conflict, could threaten our United States interests or other vital interests, and could impact our own welfare. I would use, as one example, the fisheries now in certain parts of the world. Over-fishing and ruining the reefs are causing a major depletion of resources. This destabilizing set of traumatic conditions stretch beyond just the area where they begin. In Central Command, we saw this around the southern Arabian Peninsula, the Horn of Africa, the Indian Ocean, and the Seychelles. Another example would be the increasing urbanization that we see as a result of depletion of resources out in the hinterlands. This urbanization starts the cycle of taking very fragile urban environments and drawing them down economically. They become the sources of major health problems and end up in some form of conflict.

Question: Is there a published or unpublished definition or strategy of post-Cold War security that can provide a frame of reference for the environment as a security issue?

General Wilhelm: I think we have one. Taken together, the National Security Strategy (NSS) and the National Military Strategy (NMS) provide an adequate framework at the strategic level. Environmental security issues can effectively be addressed under the "engagement" pillar of the NSS. And environmental security can be effectively addressed and even resourced under the "shaping" pillar of the NMS. That gives us the top cover, if you'll pardon the phrase, to engage meaningfully on environmental security issues. What is missing is an overarching campaign plan to operationalize that strategic guidance. Additionally, there are a couple of other things I would like to touch on. The Goldwater-Nichols Act of 1986, I think in most ways, has done an effective job of pulling the services together. It was not intended to influence our linkages with the interagency process. To me that is the next land to be conquered. One attribute that I would like to see in successor legislation would be one that would point the way toward more effective interagency coordination in areas that require civilian and military unity of effort. That, of course, includes the environment as a security and stability issue in the current global security environment.

General Zinni: [In that connection] It's not enough to just build a better sense of the relationship of the environment to security. We have to build mechanisms to deal with the problems. I feel strongly that the key problem is lack of interagency coordination and focus. The interagency process is only reactive. It only forms itself and reacts in times of crisis. Most of what we are talking about is best handled in preventive form.

Question: What role(s) do you see the military playing in an environmental security-stability scenario?

General Zinni: There are times when, in given events, the military may be the appropriate lead. At other times, particularly in the preventive phases of dealing with environmental degradation and its aftermath, civilian agencies, nongovernment organizations (NGOs), and other national and international civilian entities could and should take the lead. The answer to the general question, I think, would be situational. I think that the Horn of Africa is a good example of what I am talking about. We see both natural and man-made conditions there that impact the environment and are generating significant security concerns. Depletion of resources, water scarcity, mass migrations, urbanization, health and disease problems—just to mention a few examples throughout the region—require a coordinated and effective management approach. As a specific example, we need to work hard before we end up—before Sana'a in Yemen depletes the country for its own benefit—with a catastrophe of major proportions.

General Wilhelm: Clearly, the military has an important role to play in global stability and security. That role, however, must be based on the correct identification of a range of actions—and in proper proportions—that is subordinated to and in support of the larger political, economic, social, and security components of the environmental security-stability issue. Military leaders, planners, and implementers would do well to scrutinize seriously a long list of strategic and operational imperatives that may be derived from the linkages between environmental stressors and violence, conflict, and state failure.

Question: Would you outline a few of the most salient cases or examples of environmentally related security-stability problems that might require policy decisions sooner or later?

General Zinni: I think water is what I would put as the first priority, and all the problems related to that. Pollution, agricultural and industrial practices, upstream control and management, cultural problems in dealing with conservation—all these problems are close to being catastrophic problems in many regions of the world. Second to that is, I think, depletion of rain forests and some of the foliated areas. Third is the depletion of seabed resources. I am deeply concerned about the Eastern (China, Sri Lanka, and others) movement into the Indian Ocean and into the Gulf of Oman and around the Horn of Africa. Eastern fishing practices that are poisoning the reefs and destroying habitat have long-term effects. Sooner or later, we are going to have to help the victims of these practices. It's better done sooner than later. You can't wait until after warlordism has become the only known form of life, and rogue militias are made up of twelve and thirteen-year-old boys. We've arrived too late at that point.

General Wilhelm: There are thirty-two countries in the Southern Command's area of responsibility. If you look closely enough, I think you would find that all have one or more reasonably important environmental security issues that will probably demand some level of attention in the not-too-distant future.

Could any of these cause state failure? I think the answer to that question is that many of them probably could. Could environmental emergencies and related mismanagement cause conditions that would produce external intervention? Again, my answer to that question is yes—and we may see it very soon. Could the loss of a single crop (e.g., bananas) or industry (e.g., tourism) lead to economic collapse? The answer is yes. So, to review the bidding—what should a definition of environmental security include? My answer would be that a definition should include environmental conditions, natural or man-induced that could cause national or regional instability, state failure, external intervention, or economic collapse.

Question: Finally, what must be done?

General Wilhelm: I see no way to separate environmental security issues from the global stability problem. It is important to start working on them before whatever you do is irrelevant. For example, no football coach in the NFL would go into a season without a "philosophy" for his team. Neither would any coach in the NFL go into any given game—say nothing of the Super Bowl—without a very carefully thought-out, and thoroughly practiced, "game plan."

General Zinni: We've got to get serious about the relationship of the environment to stability and security. The problems are there. You can see them. I mean, the threats are predictable.

The contributors in this book demonstrate the argument that there is a Hobbesian calculus at work in the world today. The cumulative political, economic, social, and security costs of environmental degradation, sooner or later, will cancel out the growth from unconstrained exploitation. In the global security arena, the results are tension, instability, violence, and possible sate failure. In the absence of preventative and mitigative balancers, the international community is metaphorically eating its seed corn. Life will indeed become "nasty, brutish, and short."

The authors, then, commend this book to the reader to provoke thought about what should be done in the face of environmental degradation, violence, instability, internal and external conflict, and possible state failure. Equally important, they suggest what we as citizens ought to demand of our governments and international organizations. In that context, they commend this compendium to you—the reader—with the hope and expectation that you will make effective use of the insights of the contributors.

This book would not have been possible without the active support and assistance of several people. First and foremost, appreciation must be expressed to the contributors—Dr. Steve Blank; Vice Admiral Paul Gaffney, USN; Mrs. Darci Glass-Royal; Captain Steve Kiser, USAF; Ambassador Frank McNeil; Commander Ray Simmons, U.S. Navy (Ret.); Dr. Aondover Tarhule; and Dr. John Warren. Additionally, General Charles E. Wilhelm, General Anthony Zinni, and Ambassador Frank McNeil deserve a great deal of credit for their help, encouragement, and perceptions. Unfortunately, the contributions of Generals Wilhelm and Zinni to this project were cut short by the events and aftermath of September 11, 2001. Nevertheless, their contributions are significant—and appreciated. At the same time, our editor, Martin Hayward, made us more articulate than we really are. And, I must acknowledge—one more time—Darci Glass-Royal and Ray Simmons. Without their constant support, prodding, pulling, and pushing, this compendium would still be only an idea.

Finally, neither this volume nor the individual chapters in it should be construed as reflecting the positions of the United States government, the Department of Defense, or the Department of State. The editor and the individual authors alone are responsible for any errors or lack of judgment.

Max G. Manwaring
Carlisle, Pa.

Note

1. Author interviews with General Charles E. Wilhelm on February 9, 2001, and June 22, 2001; and with General Anthony L. Zinni on June 2, 1999, and October 6, 2000, in Reston, Va. and Washington, D.C., respectively.

INTRODUCTION

Making Sense of Environmental Security

Frank McNeil and Max G. Manwaring

This chapter is divided into two parts. Part one is an explanation of where we are today in terms of understanding whether environmental security is real and whether it matters. It derives from an essay written two years ago as an Agenda Paper of the Dante B. Facsell North-South Center at the University of Miami.[1] Part two reflects the fact that making sense of the term "environmental security" is no less of a challenge today, but—because environmental threats are so vital to United States and global security—the environmental security concept must be developed into a policy tool from which to identify and address specific risks and their implications for national security.

Part One: Is Environmental Security Real, and Does It Matter?

I [Frank McNeil] have been asked to make sense of the concept of environmental security. That is not an easy task, inasmuch as there is as yet no generally agreed-upon framework for considering the matter, even after more than two decades of discussion, which began in 1977 when Lester Brown first startled Cold Warriors with this novel proposition.[2]

Surely, the source of discord lies in the fact that environmentalists take an expansive, "planetary" view of security, while those in the traditional security community—military and civilian strategists alike—have a restrictive definition. To make matters worse, the security community, not without reason, sees the claims of environmentalists as efforts to raid military budgets to support conservation. That clash will not disappear overnight. Still, it is possible to set aside these kinds of arguments for another day and seek a practical approach to the linkages, which are quite real, between the environment and security. At root, the operational concept which informs this approach is simple. Case studies, buttressed by science where possible, are suitable vehicles not only for historical analysis, but also in forecasting future conflicts, derived in some measure from environmental change.

Case studies are nothing new, dating back at least to Thucydides. Academics have used them to illustrate historical linkages, as the authors do, between environmental insult and conflict or significant tensions. What is much harder to find, and a novel aspect of this book, is work of a prospective nature which might serve, in military terminology, as a "warning of war"—or just about as important, a warning of tensions which have significant implications for instability, and for possible state failure. This is risk analysis, forecasting potential conflicts among real people in real places derived from changes in the environment. Routinely, policy makers and foreign affairs specialists make forecasts about other sources of future conflict. So, the hard question is not whether such analysis would have practical uses, but why hasn't more of it been done.

Curiously, in the last two years, there is more recognition that environmental damage can spur conflicts and tensions. Discussions in Southeast Asia and Latin America—plus earlier experimental case studies—point to a few rules of thumb for considering the real world implications of environmental security:

- Yes, there is such a thing as environmental security. In the metaphor that environmentalists love, environmental security is a *canary in a coal mine*, warning of the danger of conflict or seriously conflictive implications.
- Environmental security is a convenient shorthand for discussing certain risks of conflict, if you think locally and in real time. Framed this way, it is clear that environmental damage has already spurred conflict in many locales.
- In the fairly near future, say, over the next five to ten years, there is every reason to believe that environmental damage will be an increasing factor in conflicts and tensions, within nations and across frontiers.

In the wake of the destruction done by Hurricane Mitch in Central America and the burning of Indonesia's forests, there is less debate about the existence of some sort of link between the environment and security concerns. Beyond this point, however, there is still no agreed-upon sense of the implications for international security policy, so that the matter remains primarily the concern of academics, nongovernmental organizations (NGOs), and international organizations (IOs).

In part, this attitude reflects the lack of an overarching concept of security to replace the one which went away with the Cold War. The terrorist events of September 11, 2001, have given us a cruel enemy, which must be fought, perhaps for years, but they also sharpened the contradiction between the world in which we want our children and grandchildren to live and the one in which we actually reside. Even so, except for terrorism, there is as yet little to concentrate the minds of policy makers in a strategic direction, so that security is still being redefined in fits and starts.

In these circumstances, environmental security, if it is to have meaning for decision makers, policy makers, and informed citizens, will have to be built from the ground up, by getting the science right and drawing from local realities inferences about specific risks and the implications for the United States and the rest of the global community. But, analysis of how all this fits together is must less advanced because the security establishment has not yet realized that the environment can kill.

So, where are we? Environmental insults—call them stresses, or stressors if you prefer—contribute to conflict and tensions within and among nations, although environmental damage is not a sole cause of conflict. Wars, after all, have many fathers. In a given locale, environmental change may not even be a principal driver, though that cannot be ruled out. Resource scarcity, for example, often drives conflict. In particular, "water wars" and lesser disputes over fresh water and maritime rights, appear to have a future, particularly in the developing world.

The conventional wisdom (if one can use that phrase about a subject in which much is controversial) holds that environmental degradation is far more likely to cause instability within nations than to feed cross-border conflicts. That conclusion, derived from an excess of theory, appears much too optimistic. In many places, discussed in the chapters that follow, environmental degradation is in fact applying stress across borders.

Thus, there is reason to believe—given the spread of environmental degradation—that the environment and security will become more tightly linked in the future. Sooner or later, U.S. policy will be driven to seek mechanisms for prevention, mitigation, and peaceful resolution of environmentally derived conflicts. This cannot be done alone, although American leadership will be required. Today, the United States, Canada, Europe, and Japan lack an agreed-upon strategy for dealing with the vulnerabilities derived from accelerating environmental damage throughout the world. To be fair, strong support exists for disaster relief, not only at home, but also abroad, where the Agency for International Development's Office of Foreign Disaster Assistance, often supported by U.S. military resources, acts as the executive agent.

Disaster response, which the public supports as a moral duty, is the only sector of America's foreign aid operations to enjoy constant congressional backing, regardless of party. One senses, however, that even these usually effective relief programs will in the long run amount to little more than triage in the absence of a more systematic address to environmental degradation.

At all events, the struggle for wise use of resources will take place—if it takes place—forest by forest, watershed by watershed, and reef by reef. Environmentalists talk about a "grand bargain" between the First and Third Worlds, which is clearly applicable to global climate change. But in the realm of environmental security, it will be necessary to "act locally," through a series of "small bargains," tailored to regional situations, to contain incipient conflicts and to help arrest the cumulative processes that sap prospects for socioeconomic growth and increase the risk of conflict.

Small bargains may not prosper in the absence of a grand strategy which persuasively links the environment, growth, and security, but talk of grand strategy will remain illusory unless environmental change first becomes widely recognized in rich and poor nations alike as a "canary in a coal mine." Here, at the intersection of security with broad policy concerns, is where the work of risk analysis lies.

Part Two: Toward Environmental Security as a Policy Tool

Whether a generalization, a pseudonym, or an oxymoron, the term environmental security is a poor label for the most likely contemporary challenge to the security interests of the United States and the rest of the global community. The term implies that this phenomenon is a singular military responsibility and requires a simple military solution. The term has also been misused, as pointed out by Frank McNeil in part one, as a means through which environmental activists might gain access to defense budgets to support planetary conservation efforts. Clearly, the United States security agenda recognizes that in a global age, threats and risks do not come only from determined enemies and deadly weapons; they may also have potential for significant impact on American interests. But, in both instances, there are distortions that need correcting.

Lessons learned from the more than one hundred conflicts that the United Nations Security Council has recognized since 1990 as serious internal and global destabilizers indicate that environmental security has meaning within the narrower context of the causes and consequences of high probability and more immediate internal conflict, rather than more nebulous planetary possibilities. The harsh realities of the new world disorder are caused by myriad destabilizers. The causes include increasing poverty, human starvation, widespread disease, and lack of political and socioeconomic justice. The consequences are seen in such forms as social violence, criminal anarchy, refugee flows, illegal drug trafficking and organized crime, extreme nationalism, irredentism, religious fundamentalism, insurgency, ethnic cleansing, and environmental devastation.

These destabilizing conditions tend to be exploited by militant nationalists, militant reformers, militant religious fundamentalists, ideologues, civil and military bureaucrats, terrorists, insurgents, warlords, rogue states, and even honest reformers working to achieve their own purposes. As a result, the interdependent global community is experiencing "wars of national debilitation, a steady run of uncivil wars sundering fragile but functioning nation states and gnawing at the well-being of stable nations."[3]

This particular theoretical approach to contemporary conflict, based on cause and effect, is heavily influenced by Thomas Homer-Dixon and Ted Robert Gurr.[4] In *Why Men Rebel*, Gurr posits the concept of Relative Deprivation (RD) and—with his associates in "The State Failure Project"—subsequently argues that the political, economic, social, and security variables that define RD may be the result of environmental stress, and can lead to internal and external violence. Importantly, environmental stressors can also contribute to state failure. Homer-Dixon, similarly, demonstrates the virtual impossibility of resolving internal instability and violence problems—and their regional spillover effects—without also dealing with the environmental causes that produce them.

Experience over time and throughout the universe of post-Cold War conflict and state failure also demonstrates that victory in any type of conflict—and overcoming failing or failed state status—is not simply the sum of the battles won over the course of a conflict, or defeating an internal adversary bent on taking down a state. Control of a given situation and its resolution are determined by two things. First, it is the product of leader judgment that connects and weights the various political-diplomatic, socioeconomic, psychological-informational, and military-security elements of national power within the context of dealing with the causes and effects of instability, conflict, and possible state failure. Second, control of a given situation and its resolution are also determined by the organization for unity of civil-military effort that is established *before*, during, and after a destabilizing problem is recognized to exist. That is to say, civilian agencies now have a role to play in a broadened definition of national security. National security is too big a problem for singular military solutions. This is a problem that requires the mobilization of all the resources and tools of government.

This brings us back to where we began—that is, to the problem of defining and operationalizing the environmental security concept. In order to get out of the intellectual vise-lock imposed by a lack of understanding and conceptualizing the phenomenon, we have chosen a practical case-study methodology to examine the linkages—which are quite real—between environment and security. Additionally, case studies also provide a suitable vehicle for forecasting tensions and future conflict. This, in essence, is risk analysis. Specifically, the case studies will:

- elaborate specific risks and threats to an economy, a society, and/or a larger community;

- provide examples of specific problems or cumulative environmental degradation acting as environmental insults—or stressors—that contribute directly to or generate internal and/or external conflict;
- demonstrate cause and effect;
- demonstrate what roles civil and/or military government agencies, international organizations, and nongovernmental organizations can play or might play in mitigating the causes as well as the consequences of environmental degradation; and
- move the issue of the environment from the stage of study and rhetoric to the realm of action—that make reasonable, specific, and actionable recommendations.

Addressing these issues at the micro and macro levels for a variety of countries and geographical areas generates, among other things, the possibilities to:

- define and operationalize a more legitimate concept of environmental security than now exists, and lift it to the rational and vital plain it deserves; and
- outline a new paradigm or overarching concept of post-Cold War security, or set of principles from which policy and strategy might flow.

Secondary objectives of the effort would be to:

- identify and address those environmental problems that will hasten, exacerbate, or even trigger instability and potential state failure, and will likely require U.S. intervention or involvement;
- focus on high risk challenges that most directly affect U.S. interests and engage other governments where we can make the most difference—in a timely manner;
- make recommendations for mitigating environmental and instability problems *before* they become crises;
- ease man-made or natural humanitarian catastrophes;
- help avert wider conflicts that might endanger global stability; and
- promote environment-related scientific research in other countries so they can better identify potential internal instability problems and develop indigenous solutions for them.

A final objective is to stimulate thinking about problems that, whether we are comfortable with them or not, or whether we are prepared do deal with them or not, are likely to be with us well into the twenty-first century. In that connection, Frank McNeil once paraphrased Harry Truman's advice—if you want results, "Scare the hell out of them." Any given chapter in this book should do exactly that.

The first chapter provides a discussion of the environmental geopolitics of the Trans-Caspian region and further introduces environmental security as a national security issue. We then offer a series of more conventional case studies that identify macro and mico level flash points around the world. The final chapter examines the questions of "What have we learned from the case studies?" and "Where do we go from here?" Taken together, all the chapters of the book answer the question posed in part one of this introduction: Environmental Security: Is It Real? Does It Matter? The answer is a resounding "Yes!"

There are no conclusions, per se. We consider this work to be a point of departure, or a commencement exercise, to move the environmental stability and security issue from the stage of rhetoric and study to the realm of action.

Notes

1. Frank McNeil, "Making Sense of Environmental Security," *The North-South Agenda* (February 2000): 1–10.

2. In 1974, with the help of the Rockefeller Brothers Fund, Lester Brown founded Worldwatch Institute, a research institute devoted to the analysis of global environmental issues.

3. Leslie Gelb, "Quelling the Teacup Wars," *Foreign Affairs* (November–December 1994): 5.

4. Ted Robert Gurr, *Why Men Rebel* (Princeton, N.J.: Princeton University Press, 1970); Ted Robert Gurr, "On the Political Consequences of Scarcity and Economic Decline," *International Studies Quarterly* (1985): 51–75; and John L. Davies and Ted Robert Gurr, eds., *Preventive Measures: Building Risk Assessment and Crisis Early Warning Systems* (Lanham, Md.: Rowman & Littlefield, 1996). Also see Thomas Homer-Dixon, "On the Threshold: Environmental Changes as Causes of Acute Conflict," *International Security* (1991): 76–116; and *Environment, Security and Violence* (Princeton, N.J.: Princeton University Press, 1999).

CHAPTER ONE

Geopolitics, National Security, and the Environment: An Example from the Trans-Caspian Region

Stephen J. Blank

Rising international acceptance of the exploitation and conservation of the environment as legitimate political issues is incontestable. The recent uproar when the United States announced it would not submit the Kyoto Protocol to the Senate for ratification signifies that environmental issues are legitimate subjects of international political debate and activity. Yet, that understanding of the environment as a subject for political action and the ability to implement meaningful and effective environmental policies vary from state to state and from region to region.

While the entire former Soviet Union was and remains an environmentally stressed area, today there is much less activism and scope for action to overcome threats to the environment in Central Asia than in the former Baltic republics.[1] Indeed, some Western observers decry the seeming political passivity of Central Asia's peoples even as Central Asia's absolutist and corrupt autocracies become ever more entrenched. Moreover, for the foreseeable future, it appears that this form of rule will persist in Central Asia regardless of the personalities involved.[2] Whereas in Northern Europe, the breakdown of social conditions in stressed regions like the Russian north and northwest or the occurrence of another Chernobyl is seen as the most urgent security threat to the region, perceptions of environmental threats to security in Central Asia are relatively few. And few scholarly analyses of Central Asian security issues are imbued with the sense of urgency that is felt by Northern European governments concerning ecological threats to security.

In fact, most scholars who analyze Central Asian security focus on interstate and foreign relations or defense and foreign economic issues, particularly as they involve the "new great game" being played over the issues of Caspian energy sources, as constituting the real issues of security.[3]

Most scholarly analyses of Central Asian security choose to use the term "security" in its more traditional sense of defense against foreign attacks or forceful internal upheaval. Consequently, analysts of Central Asia have generally "privileged" traditional concepts of security over newer concepts that place greater emphasis on soft security issues and transnational challenges, such as the environmental degradation of much of the territories of these states under Soviet rule. The primary concentration of much of this scholarly writing revolves around this new great game and the struggle for access to and influence over energy sources.

The new great game denotes the multilateral competition among states either adjacent to or outside of Central Asia for leverage regarding Central Asia's enormous oil and gas reserves. Given the economic structure of the Central Asian states, it appears that the only way those countries can modernize their economies is through the exploration, refining, transportation, and sale of oil and gas abroad. Hence, the struggle for control over these states' energy resources could determine the future economic and political structures of Central Asian governments as well as their primary orientation in foreign policy. The competition among Russia, Iran, Turkey, India, Pakistan, China, and the United States for this leverage is generally seen in the light of a classic geopolitical rivalry, particularly the Anglo-Russian rivalry over Central Asia before 1914—hence the term, new great game.

Yet, the inclination to view regional security mainly in terms of this game or of the internal stability of the various Central Asian governments is arguably incomplete. Anyone studying security issues in Central Asia quickly recognizes that environmental factors—the use and control of land, water, energy, and other raw materials, and the reclamation of polluted lands—play an extremely important role in that region's security agenda.[4]

This is as true for energy as it is for other variables. First of all, as potentially critical energy producers who depend on the income from energy for their future development, Central Asian states are naturally loathe to constrict their future opportunities by embracing binding and restrictive international environmental programs and accords. Second, the rivals for access, influence, and control of local energy sources habitually claim that their rivals' plans will endanger the fragile ecology of the region. Thus, while we cannot adduce the validity of the evidence brought forth in support of these charges in this limited space, there are well-known and generally acknowledged potential, if not actual, risks to the environment from the large-scale exploration, refining, and shipment abroad of oil and gas.[5] And it is no less the case that "Caspian ecological concerns mask political aims," even when they are well-founded.[6]

Third, even official agencies such as the U.S. Energy Information Administration (EIA) embrace the view that there is a strong connection between energy and the environment. The EIA's *International Energy Outlook 1999* report stated that, "The principal international energy issues have shifted from supply interruptions and their implications for energy security and price stability to the impact of energy production and consumption on regional and global environments."[7]

That assertion may represent a minority view; certainly, many would contest it. But it is probably not open to question that, as Central Asia increasingly finds it necessary to exploit its abundant energy resources to obtain the capital and foreign investment it desperately needs, the stresses upon the indigenous environment will grow. It is also likely that the political salience of environmental issues in these states will accompany these growing stresses. Several factors are at work here. First, since these are new states that still suffer from a shortage of environmentally and/or technically trained personnel, it is already true that much of their existing energy production is simply wasted and goes into the ground or the water table.[8] Therefore, until the government's capacity for regulating production growth develops commensurately, the pressure on clean water and arable land will probably also grow. The linked processes of development and environmental degradation will continue for some time until the local administrations have the manpower and resources to attack those problems.

Second, comparative studies of environmental issues point to the conclusion that, where the exploitation of natural resources and the directing of investment to that exploitation is of essential importance to both states and foreign investors with a direct stake in that exploitation, governments generally opt for development over environmentalism. Moreover, the dependence of Central Asian states on external sources of financing and development not only spurs the rivalry for access to oil and natural gas or other raw materials as a source of capital; it also intensifies the primacy of development over conservation. This is because the investors' prime concern is an adequate and relatively rapid return on their investment. Consequently, Central Asia's "embeddedness" in the global economy subjects it to the economic expectations and objectives of key governments, international financial institutions, and multinational corporations. This structural setting, in turn, creates the basis for conflict with local traditions and customs—which are made more acute in the absence of sufficiently trained administrators or a civil society that can mediate between the governments involved or between them and local communities.[9]

Third, Central Asian states are Third World states, and the primary concern of their governments in international economic policy making is likely to be their domestic security. That concern will take precedence over foreign policy concerns. Because Central Asian countries are Third World states, in many ways their governments display the pathologies associated with those states. For example, state intervention in the economic issues connected with the exploitation of energy or other raw materials frequently breeds an economic-political climate that favors development. It also promotes governments who view those resources as political and economic assets that they must control to remain in power. Those factors are among the classic conditions for rent-seeking economic policies to benefit current elites.[10] As the studies of Robert Bates and Jesse Ribot have observed, African governments—like Central Asian ones—use market regulations as explicit instruments of political organization in the countryside. Market intervention is also a basis for building political machines. Hence, as Bates observes, the rents that accrue to the state thereby must be seen in more than a strictly economic context of inefficient use of resources. Instead, "The prevailing tendency has been to regard these [policy generated] rents as pure social costs—inefficiencies induced by political distortions of market forces. What has not been stressed is that the rents also represent political resources—resources which can be used to organize political support."[11] Where environmental considerations restrict opportunities for using raw materials or cash crops in this way, and where domestic threats are perceived, they will likely be brushed aside in an effort to pay off potential rivals or supporters.

Indeed, it is not going too far to observe that many Central Asian states, particularly given their dependence on energy or gold, copper, and cotton as major sources of foreign currency, conform to the type of state known as "rentier states." These kinds of states live and die on the revenues extracted from the sale of natural resources. In turn, that economic necessity breeds political relationships that distort domestic economics and politics and discourage conservation policies.[12] But under such circumstances, attempts to reorient the prevailing policies pertaining to the exploitation of natural resources will involve major political conflicts. And in backward or undemocratic regimes, these contests could conceivably become violent. One need only cite the history of collectivization throughout the former Soviet Union to demonstrate the extent of the violence unleashed when a coercive state steps in to fundamentally transform long-enshrined social practices pertaining to the exploitation of the land and natural resources.

But we may also see the potential for coercive action in post-Soviet societies in a different light, especially if we factor in the current international rivalry for access to Central Asian resources. Nancy Lee Peluso observes that not only are resource-funded development strategies, such as those common in Central Asia, almost always skewed to the enrichment of the central governmental leaders; they can also foster violent struggles within or among states.

> In addition, when a state's incomplete hegemony prevents it from sufficiently controlling the people living under its jurisdiction, the state may use both conservation and economic arguments to justify the coercive exclusion of certain groups

> from valuable resources [exactly as under Soviet collectivization]. Coercion and resource control are intended to increase the state's powers of social control; and these in turn enable the state to extract more revenue from conservation or extractive zones. . . . In sum, externally based resource claimants (including the state itself) frequently redefine resources, the means by which they may be conserved or harvested, and the distribution of benefits from their protection. Such redefinitions often override, ignore, or collide with local or customary forms of resource management. When competition between external and local legitimation mechanisms is played out in the environmental arena, the result is often social and political conflict, which causes environmental degradation and ultimately fails to achieve the interests of international conservation interests.[13]

From the foregoing we can conclude the following. First, issues relating to those resources, not only energy, will directly impinge on the security considerations of local governments. Second, environmental issues relate directly to questions of governance and the distribution of assets in those societies. And those assets can include chances for better health and longevity, as well as more tangible ones such as actual possession of valuable resources and assets. Thus, environmental concerns are indeed factors that materially affect both interstate and intrastate security throughout Central Asia. The continuing debate among scholars over whether and how the environment may legitimately be regarded as part of the agenda of international security is of little practical relevance in Central Asia, because it is increasingly clear that environmental factors do contribute significantly to the regional security agenda.[14] Indeed, two prominent experts on the area, one civilian and the other military, each concluded that the most critical security threats to the area will center on environmental issues and internal security.

Water Issues

S. Frederick Starr, Director of the Central Asia-Caucasus Institute at Johns Hopkins University's Paul H. Nitze School for Advanced International Studies, writes that:

> For a region of desert and steppes, conflicts over the allocation of scarce water resources are inevitable and pose the single greatest threat to regional security. Water will become increasingly scarce in the years to come. Salinity in the Khorezm region of Uzbekistan and Turkmenistan has caused the prevailing west-east winds to sweep up tons of salt and carry it to the Tien-Shan mountains and the Pamirs. Deposited on the high mountain moraines, this salt is gradually melting the massive glaciers that are the source of water for the entire region.[15]

Soon after Starr published this assessment, General Anthony Zinni (USMC Ret.), then-commander of the United States Central Command (USCENTCOM) whose area of responsibility (AOR) included Central Asia, stated that,

"The region suffers from the horrific environmental legacy of the Soviet Union. The Department of Defense [DoD] is focusing on water-related problems in the region. Water, not energy, probably will be the cause of conflict in the region within five years."[16] Zinni's successor, General Tommy Franks (U.S. Army), also emphasized the importance of water as a factor in Central Asian security, even if his assessment was somewhat less bleak.[17]

Both Zinni and Franks have simultaneously confirmed that the key issue for U.S. policy in Central Asia is access to energy—and this also applies to the other contestants in the great game: Russia, Iran, China, Pakistan, Turkey, India, and others.[18] Questions revolving around control of that access and the exploration, refining, and transport of oil and natural gas inherently involve profoundly important issues of environmental regulation and the distribution of material assets in and from those societies. And the interest of the U.S. military—and presumably that of other militaries as well—reflects the fact that, in some quarters, the more active involvement of host country armed forces is now being called for as a way to regulate environmental questions in the Third World.[19] One can easily imagine scenarios of military conflicts within Central Asia, either among the local states, or involving foreign armed forces, over control and regulation of energy sources.

Consequently, all these warnings of potential conflicts, including those by Generals Zinni and Franks, are well founded. Indeed, there are already serious disputes over water between Kyrgyzstan and Kazakhstan.[20] And the general water situation in Central Asia appears to be worsening. In 2000 agricultural production fell by 30 percent due to limited water resources.[21] While the task of Central Asian governments arguably may be to separate national water resources from transnational ones, and to regard water basins in all the phases of the hydrologic cycle as a single natural resource to be shared equitably among all the users, this is not happening. Annual talks on water are more often bilateral rather than multilateral, and each Central Asian government declares water resources to be a "national good" so that national laws will override interregional agreements. Moreover, their media repeat this argument.[22]

Meanwhile, the desiccation of the Aral Sea is also a factor in interstate disputes in Central Asia.[23] Given the centrality of water to every form of social activity in desertlike and semi-arid climates—and the fact that water basins comprise territories of many Central Asian states, if not all of them—one can assume that Turkmenistan, Uzbekistan, and Tajikistan will also be drawn into serious environmental disputes arising out of the abuse of the environment and the ensuing resource shortages. And as the pressure of Central Asia's still rapidly growing population on depleted water supplies increases, more stress within countries and perhaps between them cannot be ruled out. In fact, this stress is probably more likely.

Prospects for Environmental Security or Conflict

Arguably, the importance of environmental issues and the urgency of the threats posed by ongoing environmental degradation in Central Asia transcends in importance the equally critical issues for Central Asian states of control over the production, refining, and export of oil and natural gas. Certainly, a movement or government oriented toward conservation as a priority might argue accordingly. It might also argue that even if the state had no oil and gas to speak of, the importance of the environmental factor would be no less apparent.

However, this argument, like so much of the environmental security debate, is also obviously a "contested" one.[24] While it is true that, "to inquire into the dynamics of global environmental change is also to probe those that sustain global social change," this does not necessarily mean that environmental issues drive or even dominate the regional agenda in Central Asia.[25] Rather, that argument poses an alternative way of looking at regional and internal security dynamics and at the concept of security itself—so the concept can include the environment as a factor in the assessment of the internal and external security of countries.[26] And this way of looking at environmental issues in Central Asia enables us to sharpen our focus regarding the region so that any study of the security agenda that does not include the environmental factor would be incomplete.

Thus, a recent Chinese assessment of the Central Asian security situation gives a very thorough list of the problems confronting the region. This assessment observes that:

> Population pressure also threatens the long-term stability of the region; according to the World Bank forecasts, by 2015 the five countries of the region will have a total population of 90 million. The geographical and climactic conditions of the region are poor, there are insufficient water resources, and there is in fact very little space suitable for economic activities, so the region can hardly bear an excessive population burden. . . .There are also serious ecological problems in Central Asia. In the past 30 years the Aral Sea has shrunk by one-half and its volume of water by two-thirds, and sea-bottom desertification has appeared over an area of 4 million hectares; as a result, abnormal climate changes have occurred, with dust and salt storms wreaking havoc, and epidemics have spread in neighboring nation, pollution from radioactive material, the exhaustion of water sources, and atmospheric pollution.[27]

This report should impart a sense of urgency to policy makers concerning Central Asia's potential for the future. Yet, when taken in tandem with the other, more immediately visible threats to Central Asian security, it becomes clear that the agenda for responding to these problems is immense and the means at hand for doing so quite limited. Thus, the report concludes that: "There are very many factors for instability, covering a very large scope, within the five countries of Central Asia, and in these circumstances any slight error can cause consequences affecting the entire situation."[28]

Accordingly, it is more accurate to observe that Central Asia's environmental problems touch on or interact with the many other issues that pertain to security there. And consideration of environmental factors helps to shape the direction of at least the policy responses of the local actors to Central Asian problems. Therefore, without favoring either traditional or newer concepts of security, we can profitably view environmental issues in this region as questions that cut across a host of other serious security challenges facing its governments. Nor is this fact confined only to Central Asia.

In the most general terms, the primary reason that so many principles and declarations of intent go unfulfilled is that environmental regimes are embedded in other social, political, and economic institutions. Although treated as such, the environment is not really a separate domain, but a crosscutting aspect of every sector. All institutions partake of environmental regimes, albeit typically of the tacit sort, in that they heave environmental consequences. Most obviously, environmental is a side effect of economic activity or, as economists call it an "externality." Certain activities that are intrinsic to contemporary life, particularly in the industrialized countries, are inherently destructive.[29]

The interactive, or crosscutting, nature of the environment as a political issue made itself felt a decade ago at the end of the Soviet period. Seventy years of Soviet lack of concern for the Union of Soviet Socialist Republic's (USSR) overall ecological balance, clean air and water, and similar resources left behind pollution problems of enormous scope throughout the entire Soviet bloc—not just the USSR or Central Asia.[30] These leftover problems include the results of large-scale experiments in nuclear power and nuclear weapons, biological and chemical warfare, industrial pollution, "harebrained" agricultural policies, and gigantically misconceived engineering projects.[31] But, more importantly, the dawning protest against that environmental degradation, whether on a republican or an all-Soviet basis, also illustrated important tendencies in political participation and new nationalist movements that emerged as the Soviet system collapsed.[32]

In Central Asia, some of the first signs of public awakening against Soviet power and arguably the first signs of a development of a civil society grew out of protests against the nuclear pollution of Kazakhstan and similar phenomena elsewhere in Central Asia. These movements were among the first to highlight the degree of environmental stress to which Central Asia had been subjected. But the subsequent development of these movements in Kazakhstan and other Central Asian states after independence was essentially thrust on them—also reflect some of the larger political dynamics of the present situation. Those developments shape the context in which any effort to organize collective or governmental action to overcome ecological threats must now function.[33]

Specifically, it became clear in Central Asia that the environmental protest was subsumed under the mantle of a generalized nationalist protest that downplayed issues of democracy and exalted the question of national sovereignty. Political action centered on environmental reform in this context could not get off the ground. And since the new leaders after 1991–92 inherited former Soviet assets and problems as well as the tendency to autocratic, corrupt, and absolutist rule, environmentalism declined and political passivity grew.[34] As the centrality of the development agenda—especially regarding energy—grew in importance, environmentalist politics fell farther into the background.

Since the advent of independence in 1991, the problems of water, energy, and land use and control over their sources have always been on the agenda—whether they facilitated regional cooperation and solidarity or aggravated regional frictions and divisions in Central Asia. Although environmental considerations are not always the prime issue at stake, they either add to regional tensions or serve as mitigating factors—depending on the context in which they appear on the agenda. But in Central Asia, these issues do not appear in a context of otherwise stable and harmonious internal and interstate relations. In fact, the situation is quite the opposite.

As Starr observes, ecological security is tied to other profoundly stressful trends that threaten the region.

> Rapid population growth, headlong urbanization and rising water consumption in the cities are depleting a water supply that is already inadequate. Urbanization is also placing social services—especially education and health care—under great strain, and financially pressed governments are incapable of mounting an adequate response. This trend, along with the growing polarization of incomes and disorientation that accompanies migration to the cities, creates potentially flammable social conditions. While a renewal of traditional Sunni Islam provides inner harmony and consolation for millions, some of the most bewildered sections of the population—both urban and rural—are attracted to politically radical forms of the faith. Corruption in both the government and private sectors provides ready justification for such discontent. Also contributing to both the corruption and the physical and moral malaise is the rapid spread of narcotics throughout the region. To date, no national or international efforts have succeeded in reducing this socially corrosive activity.[35]

At the same time, these environmental and socioeconomic tensions take place in a political environment characterized by states with low capability for domestic governance. Different analysts cite differing causes for the political manifestations of this low capability to govern effectively.[36] But the virtually unanimous conclusions of all these different assessments point to governments that possess few resources and capabilities for dealing with the multiple social, economic, and environmental challenges to stability or the tensions they generate. Thus, one study of Turkmenistan's bureaucracy offers the following, highly negative, assessment:

> Historically, there has been no Turkmenian civil service as a professional corps. Since Soviet times, recruitment, assessment, and promotion have been on an *ad hoc* ministry to ministry basis. Low, post-independence salaries and the resultant corruption have affected civil servants' professionalism negatively, caused them to have a bad reputation, and has made coherent policy toward them difficult. That is not to say that the Turkmenian bureaucracy is altogether incapable. Informal, social relations among officials enable the system to function. However, what technical capability Turkmenistan's civil service possesses pertains to fulfilling centrally planned goals and implementing the Communist party [or now Niyazov's] line. It has no experience with either democratic government or free enterprise. The country thus has some highly skilled officials, but they lack knowledge in such areas as economic and financial management, human resources, and legal and organizational development. Hence, creating a bureaucracy to support self-sustaining institutions for collective decision-making and efficient resource allocation poses a particularly daunting task.[37]

This evaluation could also apply across the board to the other Central Asian and post-Soviet states in the former Soviet Union. Indeed, it applies to a considerable degree to Asian states in general. Thus, a recent paper composed for the Asian Development Bank regarding Asian enforcement of environmental protection observes that:

> Progress in addressing many of the region's most pressing environmental challenges has been blocked by failures of governance. In the last decade, most Asian governments have developed an impressive array of environmental legislation and regulatory frameworks. However, enforcement is weak. Environmental agencies are marginalized with respect to ministries charged with promoting economic growth. Moreover, in many countries, weak civil societies and justice systems fail to provide alternative mechanisms to hold government and corporate actors accountable for their environmental performance. As a result, the pollution of air and water from household and industrial sources threatens the health and well-being of hundreds of millions of people in the region.[38]

Since 1991, no Central Asian state has been overthrown. For the most part, Central Asian governments have conducted their foreign affairs quite insightfully. Thus today, the "new great game" actually replicates a true version of multipolarity, as multiple contenders for influence compete to make their leverage felt and the competition nullifies any one power's chances for dominance. However, the juggling act of the Central Asian governments may soon draw to a close. It is now apparent that insurgency—whether over contested drug routes or as a rallying cry of the dispossessed, disenfranchised, and radicalized Islamic constituency—will be an annual and enduring fact of life in Central Asia.

Tajikistan, Kyrgyzstan, and Uzbekistan, in particular, are threatened by these insurgencies—whether they are domestic in origin or fomented entirely by Pakistan and Afghanistan—and cannot stop these insurgent radical or criminalized movements by their own means. Therefore, they are under great pressure to bring in Russian and now possibly Chinese forces who will not leave anytime soon and are themselves notoriously corrupt, brutal, and not overly concerned with supporting democratic and human rights in Central Asia. The recent decisions of the Shanghai Cooperation Organization, as it is now called, make the possibility of joint Sino-Russian forces in Central Asia a real possibility.[39] And these insurgencies, combined with growing external pressures on Central Asian governments, occur in an environment in which the challenges posed by the ecological and other issues cited above are causing increasing polarization among these governments. This polarization naturally creates more obstacles to their mutual and multilateral cooperation.[40]

Furthermore, if Africa is any guide, to the extent that these insurgents can gain control over key energy routes, pipelines, or sites of extraction, they may be able to make deals with foreign businesses or governments that will give them enough supplies and establish a relationship of mutual need and advantage that can keep them going for years. In that event, although the results may vary in pace and nature from state to state, we could expect the kinds of conflicts and outcomes that have torn apart states such as Sierra Leone or Angola.[41]

This line of analysis appears to conform to other research on the connection between issues of environmental policy and possible violent conflict within or among states. While it is not the environmental issue per se in any given state that provokes conflict, it often appears to be the case that the socioeconomic and political consequences of ecological degradation, resource shortages, and the like interact with preexisting structural social, economic, political, and other factors to make conflict more likely. While we cannot definitively trace a direct link or line of causation from environmental problems to war and violence, many ongoing and parallel studies suggest that where severe degradation occurs, conflict becomes more likely.

For example, the Environment and Conflict Project (ENCOP), conducted by the Swiss Federal Institute of Technology in Zurich and the Swiss Peace Foundation, found that, according to project director Guenther Baechler:

> Developing and transitional societies or, more precisely expressed, marginalized areas in these countries are affected by an interplay among environmental degradation, social erosion, and violence that intensifies crises. Crisis areas prone to conflict are found in arid and semi-arid ecoregions, in mountain areas with highland-lowland interaction, areas with river basins sub-divided by state boundaries, zones degraded by mining and dams, in the tropical forest belt, and around expanding urban centers. Historically situated, culturally bound societal relationships to nature are

> subjected to upheaval and put acutely at risk in subregions of Africa, Latin America, Central and Southeast Asia and Oceania.[42]

In keeping with these findings, much contemporary research suggests that geography is a major contributing factor to the continuing poverty of landlocked states such as those of Central Asia that are far from major transportation outlets. In this case, the cure for poverty and the accompanying risk of social conflict are not so much better market institutions and economic governance as large-scale investment in transportation infrastructures.[43] While such programs are under way or under contemplation in Central Asia and are vital to the region's future, they may not be numerous or large enough to overcome the active security challenges confronting the region anytime soon.[44] And, in any case, the scope of the necessary projects inevitably entails further environmental dislocation at best, if not degradation at worst.

Funding for investment in those projects must come from abroad and will inevitably entail greater or lesser restrictive political conditions. Second, the success of these projects, e.g., the joint Russo-Iranian-Indian project or the European Union's Silk Road project, also depends on the ability of these projects to link Central Asia with major ports, airfields, and rail terminals in other countries. Therefore, these projects become a source of complex political struggles over priorities and the international economic orientation of these states. The same holds true for energy deals and pipelines. In other words, to obtain the capacity to deal effectively with looming and preexisting environmental threats, Central Asian governments are forced to risk even more degradation and its consequences. Faced with such difficult choices, the natural reaction of these governments is to temporize, muddle through, or "satisfice" the demands imposed on them. And since Central Asian states are governments that are weaker than most of their interlocutors in world politics and economics, their ingrained foreign policy reflex is to diversify choices and place many bets on many options—not necessarily the correct economic strategy in this context.

At the same time, these governments must respond to present emergencies and crises. Because they are also autocratic and authoritarian, with low capacity for governance, and few constraints on the rulers' personal aggrandizement are discernible, the persistence of civil strife will place further pressure on the region's environmental burdens in several ways. Obviously, funding that alleviates some of the problems of water, desiccation of the Aral Sea, nuclear and other pollution, and pollution resulting from intensified energy exploration and production will be diverted to internal and external security. Likewise, the priorities of policy makers will shift to those issues rather than to more developmentally based agendas that alone can improve people's lives and overcome crushing poverty. But beyond that, we can surely expect that ongoing warfare will further destroy the vital resources needed to construct viable environmental regimes and will impede the development of governmental capabilities to administer any solutions that might otherwise be devised in response to environmental issues—be they individual, bilateral, or multilateral.

Therefore, while it would be rash to ascribe pride of place to environmental issues as a factor challenging Central Asian security, their occurrence in a context where other, more traditional security challenges are rife and growing, heightens the stresses on local governments and peoples. This finding conforms to earlier research on conflicts over water, which tends to confirm the warnings issued above by Zinni, Starr, and Franks.[45] Thus, Stuart Horsman writes that, "riverine water, particularly when linked with irrigated land, is perhaps the only regional environmental issue that demonstrates a 'probable linkage between environmental degradation and the outbreak of violent civil or interstate conflict.'"[46]

Horsman's analysis of the water issues in Central Asia shows that they are clearly linked to vital national interests and to high-level political adjudication or resolution.[47] While the context in which states address these issues varies with the political and social contexts, in a number of cases contested water issues add to an already unfavorable political and socioeconomic relationship between Central Asian states or between a Central Asian state and one or more of its external neighbors.

In the first instance, Horsman cites the tense relationship among various bilateral pairs of Central Asian states, with the most contentious one being between Uzbekistan and Turkmenistan. As he notes:

> Water disputes also pervade relations between Turkmenistan and Uzbekistan. For a number of qualitative and quantitative reasons . . . the tense personal relationship between Presidents Karimov and Niyazov, the economic significance of irrigation agriculture and the two states' near-total dependence on the Amu Darya . . . water has been elevated from a political dispute to a potential cause of conflict. . . . [Thus] [t]he Turkmenistan-Uzbekistan situation may be the single most serious regional resource-related dispute. Border and water disputes have raised stress levels between the capitals. Military tensions among the Bukhara-Lebap border were recorded in 1995. Water may have been the source of this incident. A Russian newspaper reported that Uzbekistan has contingency plans for the occupation of north-eastern Turkmenistan. Rumors of these alleged plans do little to establish confidence between the two states.[48]

At the same time as bilateral tensions exist between states, certain regions within Central Asian states—because of their water situation—offer a high potential for intrastate war. Among them is the Batken area of Kyrgyzstan, which is now the site of repeated conflicts between the allegedly Islamicist Muslim insurgents coming out of Afghanistan and the government.[49]

Horsman's second kind of conflict, between Central Asian states and their external neighbors, also could present serious security challenges in the future. Since Afghanistan shares the Amu Darya and other key rivers with Central Asian states, any postconflict effort to develop Afghanistan and restore hydroelectric projects could, in the absence of consultation with downstream states, lead to disputes and even further conflicts. Existing evidence already suggests the makings of a tense atmosphere of charges and countercharges with respect to Afghan water policies.[50]

Even more suggestive is the case of China's large-scale water projects for Xinjiang, which could affect Kazakstan and Russia quite significantly. China plans to extract water from the Ili and Irtysh Rivers for Urumchi and oil field development in the Xinjiang autonomous Uighur Region—a source of escalating interethnic conflict between Han and Uighur or between Muslim and Chinese. While both rivers rise in China, the Ili passes through Kazakhstan before terminating in Lake Balkhash, and the Irtysh River travels through Kazakhstan before joining up with the Russian Ob River and Siberia. This proposal aims to stimulate Xinjiang's economy, while eroding support for Uighur unrest or what Horsman calls "irredentism."[51] But it will probably fail to meet its goals, while at the same time linking "environmental degradation and political activism, erosion of regime legitimacy, and instability."

Thus, this project will also lead to outcomes similar to those we now see in Central Asia. Xinjiang is already apparently the most environmentally stressed area in China.[52] This project will also have negative impact on Kazakhstan, which is already involved in a host of water disputes with other neighbors in Central Asia, and other environmental concerns relating to oil and gas and Soviet biological and chemical warfare experiments. While Kazakhstan alone cannot really challenge China, if the context of Sino-Russian relations changes, the deleterious impact of this policy on Russian conditions as well could become a factor in the linking up of hostile Kazakh and Russian demands upon China.[53]

Here again, there is good reason for concern about the future ecological consequences of Chinese policy, due to China's high-handed attitude. China has contravened both international law and bilateral agreements specifically by not notifying Kazakhstan of its intentions and not providing environmental impact assessments. Thus, "it is evident that China is unwilling to engage in meaningful cooperation or compromise [in] the pursuit of its water demands."[54]

On a broader scale, China's policy resembles that of other Asian states, which are determined to ensure the security of their national water supplies—often at the expense of the interests and stability of other states and regions. Therefore,

> Acute water scarcity in Asia, competitive demands for water, exclusive perceptions of water rights and complex diplomatic relations make cooperative water management "increasingly improbable." In essence, this security dynamic will be deter-

mined by the culture of regional diplomacy, traditional interstate alliances, and geopolitical concerns as much as by the core water dispute.[55]

Unfortunately, China's arrogant unilateralism feeds into Kazakh popular and elite anti-Chinese sentiments, adding an environmental dimension to them. It could also further invigorate political activism on behalf of China's rebellious Uighurs.[56]

China's attitude seems to color much of its foreign policy toward Central Asia (as well as other states) and reflects what has been called "the middle kingdom complex." China's posture vis-à-vis Central Asia is clearly geopolitically determined, if not overdetermined. According to one recent study, China looks at Central Asia, first of all, on the basis of the strength of the five individual Central Asian states, their capacity for cooperation, and their stance in world politics—i.e., their ability to form alliances with either Russia, China, or the United States. That capacity then adversely affects the party who is left out of the system.[57]

The second perspective for Beijing is its consuming interest in the region's stability and prosperity in order to extinguish or reduce support for the insurgents in Xinjiang.[58] Thus, "economic development is a central component of China's approach to fighting separatism and maintaining long-term stability in Xinjiang."[59] If the environmental interests of Kazakhstan are adversely affected thereby, China views this as unfortunate, but also too bad. Despite a rhetoric that professes to shun spheres of influence, military blocs, and the like, it is clear China and Russia have formed such a bloc in Central Asia, despite their own economic and political "contradictions." And it is equally clear that this bloc is squarely aimed against the United States. Moreover, as in other areas, China's rhetoric—though clothed in velvet—betrays a government that is more than ready to come out fighting on behalf of great-power unilateralism, an inherently uncooperative approach to security. Hence,

> China is not against the Central Asian states developing military cooperation with other countries. It supports them in their task of developing their own armies and defence systems, and welcomes the efforts of relevant countries and international organizations to support their peace and stability. However, China retains its own principles and security interests. In its view the involvement of large countries should not allow an arms race or military confrontation to develop between the states of Central Asia, and other countries and external forces should not be able to intervene in the internal politics of the states in the region or harm their independence and sovereignty. China will neither seek to obtain a sphere of influence among the Central Asian states nor force its will and ideas on these states. *It does not, however, allow activities to take place in the Central Asian states that harm its own sovereignty or are directed at breaking up China. Any cooperation should be aimed at serving the interests of peace and the development of Central Asia and the whole of Asia.*[60] [Emphasis added]

China's truculence on water issues fits perfectly with this patronizing and explicitly neocolonialist perspective.

Conclusions

This chapter has not attempted to provide an exhaustive taxonomy or analysis of existing environmental issues in Central Asia and their impact upon security—neither in the classical meaning of the concept nor its newer definition.[61] Rather, I have sought to introduce the reader to the fact that major ecological or environmental issues affecting both the internal and external security of Central Asian states—either individually or as a region—exist. And these issues are likely to become more serious and have a more serious impact in the future. As Starr observes,

> Analyses of strategic issues in the Caspian Basin focus mainly on the geopolitical situation and only touch incidentally on internal factors. However, if the most dangerous scenarios are those that would be triggered by domestic breakdown within the region, then internal factors must be accorded graver attention than they normally receive.[62]

Thus, the most dangerous potential crisis facing Central Asia is one in which the outcomes generated by the internal factors—noted by Starr and the Chinese report cited above—interact with the external threats of violence, insurgency, and the Russo-Chinese pressures for joint counterinsurgency operations against the rebels or for economic warfare in connection with energy issues. Since environmental pretexts or realities can, as noted above, conceal political or military agendas, it is conceivable that ecological issues could figure very prominently in either military scenario in Central Asia or in major international struggles over energy exploration, reigning, and pipelines.

There is already a sizable body of research into the linkage between the environment and the challenges to security in Central Asia more generally, and studies of the specific issues in Central Asia. These studies are available to anyone who is interested in the subject. But even this introduction points to the following conclusions. First, Central Asia's states entered the world as ecologically stressed areas, and their ecological situation has probably worsened since 1991–92. Second, the challenges to ecological security involve fundamental issues such as energy and water, and thus exercise a significant, if varying, impact upon the relationships of Central Asian governments with each other and their neighbors. Third, it is still unclear if these governments can muster the necessary domestic and administrative resources to begin to overcome these challenges. Since this region is already subjected to numerous contending external power influences and rivalries, yet remains dependent—for all intents and purposes—mainly on one or two "cash crops" or natural resources such as oil and gas, it will also be subjected to intense external pressures for development at all costs. Consequently, the interplay between internal and external pressures on its governments will increase. And, in view of the conclusions listed above, so, too, will the factors making for environmental degradation increase. Thus pressure on the environment in Central Asia in the forseeable future will likely increase considerably. The factors that are leading to this rising pressure on the environment include international and internal pressure for accelerated development of energy sources, population growth, and pressure on water and land as well as the security interests of neighboring states inside and outside Central Asia with regard to issues of water, land, and energy.

Since poverty pushes the local governments toward a developmentalist imperative that privileges development at the expense of conservation and an ecologically oriented policy, the area faces a paradox. In order to accumulate the resources and capabilities needed to confront environmental issues, its governments must follow policies that further stress the environment and increase the risks to security. This is an unenviable prescription for the future. But in world politics, as in life, beggars cannot be choosers.

Notes

1. Murray Feshbach and Alfred Friendly, *Ecocide in the USSR: Health and Nature under Siege* (New York: Basic Books, 1993); Murray Feshbach, *Ecological Disaster: Cleaning up the Hidden Legacy of the Soviet Regime* (New York: Century Foundation, 1995).

2. "Democracy in Central Asia; Don't Know, Don't Care in Kazakstan," *Economist* (June 9, 2001), 40.

3. The recently edited book by Roy Allison and Lena Jonson, *Central Asian Security: The New International Context* (London: Royal Institute of International Affairs and Brookings Institution Press, 2001), is an exception in this regard; see also S. Frederick Starr, *The Security Environment of Central Asia* (Emirates Center for Strategic Studies and Research, Emirates Lecture Series, Abu Dhabi, United Arab Emirates, 1999), 4.

4. Robert Ebel et al., *The Geopolitics of Energy into the Twenty-first Century* (Washington, D.C.: Center for Strategic and International Studies, 3 volumes, 2000).

5. Thus, one reason Turkey is adamant about supporting a pipeline for Central Asian energy that goes from Baku to its Mediterranean port of Ceyhan is its professed fear, based on actual incidents, that a major expansion of tanker traffic through the Straits of Marmara greatly heightens the risk of an environmental disaster affecting Istanbul's 10 million people, National Public Radio, *All Things Considered*, June 17, 2001.

6. Michael Lelyveld, "Caspian: Ecological Concerns Mask Political Aims," *Times of Central Asia* (June 11, 2001); "Oil Pipelines Pose No Threat to Caspian Environment Says U.S. Official," *Agence France Presse* (June 7, 2001).

7. Ebel, et al., vol. 3, p. 9, quoted from *International Energy Outlook 1999* (Washington, D.C.: Energy Information Administration, U.S. Department of Energy, March 1999).

8. Radio Free Europe/Radio Liberty, *(UN) Civil Societies* (June 6, 2001).

9. Nancy Lee Peluso, "Coercing Conservation: The Politics of State Resource Control," in Ronnie D. Lipschutz and Ken Conca, eds., *The State and Social Power in Global Environmental Politics* (New York: Columbia University Press, 1993), 50–52.

10. For the concept of rentier states, see Jill Crystal, *Oil and Politics in the Gulf: Rulers and Merchants in Kuwait and Qatar* (Cambridge: Cambridge University Press, 1990); Giacomo Luciani, "The Oil Rent, the Fiscal Crisis of the State and Democratization," Ghassan Salame, ed., *Democracy without Democrats: The Renewal of Politics in the Muslim World* (London: I.B. Tauris, 1994), 130–55; Jacques Delacroix, "The Distributive State in the World System," *Studies in Comparative International Development*, 15, no. 3 (fall 1980): 3–21; Hazem Bebalwi and Giacomo Luciani, eds., *The Rentier State*, vol. 2 (London: Croon Helm, 1990); H. Mahdawy, "The Problems and Patterns of Economic Development in Rentier States: The Case of Iran," M. A. Cook, ed., *Studies in the Economic History of the Middle East* (London: Oxford University Press, 1970), 428–67.

11. Jesse Ribot, "Market-State Relations and Environmental Policy: Limits of State Capacity in Senegal," in Lipschutz and Conca, eds., 34; Robert H. Bates, *Markets and States in Tropical Africa: The Political Basis of Agricultural Policies* (Berkeley: University of California Press, 1981), 99–100.

12. See note 8 and Stephen J. Blank, "Russian Democracy: From the Future to the Past," *Demokratizatsiia* 6, no. 3 (summer 1998): 550–77; and "Every Shark East of Suez: Great Power Interests, Policies, and Tactics in the Transcaspian Energy Wars," *Central Asian Survey* 18, no. 2 (1999): 149–84.

13. Peluso, "Coercing Conservation," 50–51.

14. Gregory D. Foster, "Environmental Security: The Search for Strategic Legitimacy, *Armed Forces and Society* 27, no. 3 (spring 2001): 373–95.

15. Starr, *Security Environment*, 9.

16. General Anthony Zinni (USMC), *Keynote Address*, National Intelligence Council, *Central Asia and the South Caucasus: Reorientations, Internal Transitions, and Strategic Dynamics, Conference Report* (Washington, D.C.: National Intelligence Council, 2000), 10 (henceforth NIC Report).

17. *Statement of General Tommy R. Franks, Commander-in-Chief U.S. Central Command*, March 28, 2001, to the House Armed Services Committee, 107th Congress, March 28, 2001, *www.house.gov/hasc/openingstatementsandpressreleases/107thcongress/01-03-28*.

18. Franks, *Statement*; Anthony Zinni, "Avoid a Military Showdown with Iraq," *Middle East Quarterly* 5, no. 3 (fall 1998): 64.

19. Peluso, "Coercing Conservation," 66–67.

20. Almaty, *Kazakh Commercial TV* in Russian, August 10, 2000, *Foreign Broadcast Information Service Central Eurasia* (henceforth *FBIS SOV*), August 10, 2000.

21. Iskander Abdullaev, "The Central Asian 'Water Pact'," *Times of Central Asia*, May 7, 2001.

22. Abdullaev, "Water Pact."

23. John C. K. Daly, "Global Implications of Aral Sea Desiccation," *Central Asia-Caucasus Analyst*, November 8, 2000; Daphne Billiouri, "The International Response to the Aral Sea," *Eurasia net.org*, January 12, 2001; Douglas Frantz, "Despair Looks Like a Sea That Died," *New York Times*, March 4, 2001; Erika Weinthal, *In Focus: Central Asia:*

Aral Sea Problem 5, no. 6, p. 3, March 2000, www.cc.columbia.edu/sec/dlc/ciao/pbei/fpif/wee01.html.

24. Foster, "Environmental Security," 373–95.

25. James Rosenau, "Environmental Challenges in a Turbulent World," Lipschutz and Conca, eds., 79.

26. This is one of the basic themes of Lipschutz and Conca, but it is common throughout a wide range of literature on international relations; Barry Buzan, Ole Waever, and Jaap de Wilde, *Security: A New Framework for Analysis* (Boulder, Colo.: Lynne Rienner, 1998).

27. Beijing, *Xinhua Hong Kong Service in Chinese*, September 19, 2000, *Foreign Broadcast Information Service China* (henceforth *FBIS CHI*), September 19, 2000.

28. *FBIS CHI*, September 19, 2000.

29. Karen Litfin, "Regimes: Playing Tug of War with the Nation-State," in Lipschutz and Conca, 107.

30. Feshbach and Friendly, *Ecocide*; Feshbach, *Ecological Diaster*.

31. For example, Almaty, *Interfax-Kazakhstani, in Russian*, August 7, 2000, *FBIS SOV*, August 7, 2000; Judith Miller, "At Bleak Asian Site, Killer Germs Survive," *New York Times*, June 2, 1999, 1.

32. Barbara Jancar, "The Environmental Attractor in the Former USSR: Ecology and Regional Change," in Lipschutz and Conca, eds., 168–79.

33. Jancar, "The Environmental Attractor."

34. Jancar, "The Environmental Attractor."

35. Starr, *Security Environment*, 29.

36. E.g., NIC Report.

37. Robert C. Rickards, "Business, Bureaucrats, and the EU in Turkmenistan," Frank Columbus, ed., *Asian Economic and Political Issues* (Commack, N.Y.: Nova Science Publishers, Inc., 1999), vol. 2, 212.

38. Lyuba Zarsky, *From Bystanders to Collaborators: New Roles for Civil Society in Urban-Industrial Environmental Governance in Asia*, prepared for the Working Paper Series accompanying the Asian Development Bank's *Asian Environment Outlook 2001*. Available from www.nautilus.org.

39. Robert Karniol, "Shanghai-Five in Major Revamp," *Jane's Defence Weekly*, June 27, 2001, 5; Bates Gill, "Shanghai-Five: An Attempt to Counter US Influence in Asia?" *Newsweek Korea*, May 2001, www.brookings.edu/views/op-ed/gill/20010504.htm.

40. Martha Brill Olcott, "Central Asia: Common Legacies and Conflict," in Allison and Jonson, eds., 24–47.

41. Ebel et al., *Geopolitics*, vol. 2, 55–58; William Reno, "Internal Wars, Private Enterprise, and the Shift in Strong State-Weak State Relations," *International Politics*, 37, no. 1 (March 2000): 57–74; Idem., *Warlord Politics and African States* (Boulder, Colo: Lynne Rienner, 1999).

42. Guenther Baechler, "Environmental Degradation in the South as a Cause of Armed Conflict," Alexander Carius and Kurt M. Lietzmann, eds., *Environmental Change and Security: A European Perspective* (Berlin: Springer Verlag, 1999), 108, cited in Anne Ehrlich, Peter Gleick, and Ken Conca, "Resources and Environmental Degradation as a Source of Conflict," Paper presented to the 50th Pugwash Conference on Science and

World Affairs: "Eliminating the Causes of War," August 3–8, 2000 (Cambridge, United Kingdom: Queens' College), www.cc.columbia.edu/sec/dlc/ciao/conf/eha01/eha01.html.

43. Ricardo Haussmann, "Prisoners of Geography," *Foreign Policy* (January–February, 2000): 44–53.

44. On the strategic significance of roads and means of transport here, see Mahnaz Z. Ispahani, *Roads and Rivals: The Political Uses of Access in the Borderlands of Asia* (Ithaca, N.Y.: Cornell University Press, 1989).

45. NIC Report, 10; Starr, *Security Environment*; Franks, *Statement*.

46. Stuart Horsman, "Water in Central Asia: Regional Cooperation or Conflict," in Allison and Jonson, 69.

47. Horsman, "Water in Central Asia," 69–94.

48. Horsman, "Water in Central Asia," 76.

49. Horsman, "Water in Central Asia," 74–77.

50. Horsman, "Water in Central Asia," 78–79.

51. Horsman, "Water in Central Asia," 79–80.

52. Horsman, "Water in Central Asia," 80–81.

53. Horsman, "Water in Central Asia," 81.

54. Horsman, "Water in Central Asia."

55. Horsman, "Water in Central Asia."

56. Horsman, "Water in Central Asia."

57. Guancheng Xing, "China and Central Asia," in Allison and Johnson, 152–53.

58. Guancheng Xing, "China and Central Asia."

59. Guancheng Xing, "China and Central Asia," 164.

60. Guancheng Xing, "China and Central Asia," 168.

61. Buzan, Waever, and de Wilde, *Security*; Lipschutz and Conca, *State and Social Power*.

62. Starr, *Security Environment*, 4.

CHAPTER TWO

Security Implications of Asia's Environmental Problems

Frank McNeil

This chapter really began in 1979, when I was in the Philippines for six weeks on a mission for the State Department. Even then, well before the overthrow of the Marcos dictatorship in 1986, knowledgeable Filipino, Southeast Asian, and American officials recognized that environmental loss—deforestation, in this instance—was contributing to political instability and violence.[1]

The Philippines

Richard Holbrooke, then Assistant Secretary of State for East Asian Affairs, invited me to join him on a small Air Force "attaché plane" for a look at U.S. bases at Clark Field and Subic Bay and their environs, a flight which took us over central Luzon. We knew, of course, that the forests had taken hard blows. Still, on a clear day, it was startling to see the destruction from the air. For Holbrooke and me, that flight over Luzon became a defining moment.

The only big forest in sight, except for the upper reaches of Mt. Pinatubo, later damaged by that volcano's huge eruption, was the mountain watershed above Subic Bay and the airfield at Cubi Point, kept in a state of nature by guards from the base. The uncultivated flatlands were given over to grasses and the hills to scrub brush, even well away from the towns.

Back then, Filipino and American observers usually cited the island of Negros as the place where deforestation had sparked political trouble. Clearing forests for sugar plantations, owned by Marcos' cronies, had severely disrupted the rural and village way of life. The regime had pushed sugar—then a cash cow for Marcos and his cronies—without regard for the consequences. For some, there was no place to go but the slums of Manila. Others stayed, thrust into a semirural proletariat, as low-paid plantation workers. Wherever they went, they took their resentments with them and passed them on to their children.

As we had seen for ourselves, the devastation was not limited to Negros. Of the major islands, only Mindanao in the south remained largely untouched. Elsewhere, loggers, propelled by the international demand for hardwoods and luan logs for pulp, were stripping much of the old growth, a process that would continue unabated until Marcos' ouster, by which time much of the Philippines north of Mindanao had lost much of its old growth forest.

In the once lush province of Batangas, which I visited in the company of its governor, even wood for cooking had grown scarce. He had begun a serious effort at reforestation with an import, the *ipil ipil* (a newly planted tree that could reach nearly twenty feet in a year or so), for firewood and fast growing shade, particularly along the roads and in the towns. Today, we know that reforestation with native species is preferable, but no matter; the governor was ahead of his time.

Few Filipinos lived in the heart of the forests, but relentless cutting had disrupted the way of life and the work patterns of populations on their edges. Poor to begin with, they lost the structure which traditional rural society had given to their lives: numbers of the dislocated from around the Philippines found their champions in the communist guerrillas of the New People's Army (the NPA). Their youth were the recruiting pool for the NPA.

In the late 1980s, an Austrialian public relations outfit made a video for a Japanese multinational with extensive logging interests (names omitted to protect the guilty). The video, filmed in the Philippines for English-speaking as well as Japanese audiences, contrasted the former clearcutting of entire hillsides with the firm's new method, bulldozing roads into the heart of the forest and taking the big trees (in the video, all of them) out in big trucks. The announcer, a cheerful sort, made much of the employment generated during the logging, but neglected to say what happened to the Filipino workers after the logging had ended. He said the abandoned roads returned to a state of nature, but the video, taken two or three years afterwards, showed a road's conversion into gullies, carved out of the topsoil by the annual heavy rains.

Flash back to the 1950s. In the aftermath of Philippine independence, the Huk guerrillas, who enjoyed support from Moscow and Beijing, sought to overthrow the government, only to be defeated by a popular elected president, Ramon Magsaysay. The United States recognized the societal roots of the conflict and foreign aid was used to bolster Magsaysay's efforts at land reform and rural development to "win the hearts and minds of the people."

It is outside the scope of this chapter, but worth noting that the Phillipines and a parallel British effort which defeated a similar insurgency in Malaysia, during the last stages of British rule, provided a pattern for later American counterinsurgency efforts: It was in the Philippines, for example, that Edwin Lonsdale, the model for the hero of the novel *The Ugly American*, learned the lessons that failed him in a later and different war, that of Vietnam.

These lessons were ignored by the Marcos regime, and in the 1970s, the NPA constituted itself from the remnants of the Huks, financing itself through depredations of various sorts. Though the U. S. government, until the Carter administration, gave uncritical support to Marcos, early on U.S. officials in Manila and in Washington identified social disruption and urbanization run amok as causes of popular discontent, together with repression and mounting corruption. In probing deeper, they recognized that deforestation, through the social disruptions it created, was a significant factor in the conditions that allowed the NPA to grow and prosper.

By 1986, the NPA threatened the regime, in what had become a textbook guerrilla war. Marcos was in a state of denial, disputing the threat even as he regularly asked for more military help from the United States, which was providing substantial aid and training to the Philippine military. Some army units fought aggressively, but many were reduced to garrison duty in the towns. They might go into the countryside by day, but as U.S. military observers said, the night belonged to the NPA. The NPA was exceedingly cruel, prompting comparisons with Cambodia's Pol Pot should the insurgents win. Even so, popular resentment with the Marcos regime provided the sea, in Mao's adage, in which the guerrillas swam comfortably.

It is relevant at this point to note that Holbrooke came to Manila in 1978 on a mission to pressure Marcos to release Ninoy Aquino, the leader of the democratic opposition, from a trumped-up death sentence. Marcos freed Aquino to go to the United States. By the time I returned to Manila in 1986, I was headed for meetings with, among others, his widow, Corazon. Murdered by Marcos' thugs at the Manila airport, in death Aquino became the spark for the astonishing events which culminated with Corazon Aquino's election to the presidency and Marcos' ouster.

Though Marcos still had defenders in the American establishment, notably the Central Intelligence Agency's William J. Casey, by this time most American observers, official and nonofficial, had concluded that it was only a matter of time, two or three years at most, before the NPA would oust the Marcos regime. This judgment, shared by many in the U.S. Congress, formed the basis for a major policy change. The Reagan administration, which had supported Marcos as a bulwark against communism and the guarantor of U.S. bases in the Philippines, recognized the legitimacy of Mrs. Aquino's election, which Marcos had sought to steal with his customary clumsiness. After no little difficulty, the United States flew Marcos to exile in Hawaii.

Neither indiscriminate logging nor the clearing of forests for sugarcane created the NPA, nor were they the prime causes for Marcos' downfall, which lay in oppression and staggering corruption (the term "kleptocracy" was coined in honor of the Marcos regime, though it has application elsewhere.) But the destruction of the forests helped create social conditions in which the defeated Huks could resurrect themselves as the NPA and become, until Marcos' ouster deprived them of their cause, a successful insurgency.

The Marcos era is an archetype, a nearly perfect example of how large-scale environmental change feeds instability and violence. This history speaks directly to a national security and foreign affairs community which remains sceptical of the relevance of environmental security to "real world" security concerns. In fact it was precisely "real world" considerations, the construction and preservation of a favorable Cold War balance of power in Asia, which kept the Philippines in the forefront of U.S. government attention.

In the literature, the Philippines (along with Haiti, with its own set of "real world" considerations, related to immigration) is the most frequently cited example of how large-scale environmental change can contribute to instability and violence. Hindsight is easy, and often accurate, in the hands of those skilled in historical analysis. What makes the Phillipine case even more interesting in the context of comparative analysis is that *contemporary government analysis* had discerned the linkage.

Without knowing it, U.S. officials had invented a practical use for environmental security. With the Philippine experience as a guide, it should have been a natural progression for officialdom to have added large-scale environmental changes to their template of indicators of conflict. But this hasn't happened, not yet, at any rate. Lessons were quickly drawn about the relevance of democracy and human rights to political stability, but even today the national security community pays very little attention to environmental concerns.

Of course, the Philippine case will not repeat itself; the singular personas of Ferdinand and Imelda Marcos are beyond the inventive powers of a novelist. But the Phillipines represents a pattern, evident in the tropics, where the loss of environmental assets can become a significant factor in the deterioration of a society. Perhaps it is not too much to ask of our officials to remember the Philippines and recognize that in the "real world," environmental change can be a "canary in a coal mine," an indicator of tension within and among nations.

Afterthoughts

The expectation of Filipinos and Americans that restoration of democracy would bring an end to the NPA insurgency proved to be right. The Abu Sayef insurgency hangs on in the south but is isolated from the rest of this large country. The Philippine body politic has had more than enough of insurgency and terrorism, one of several reasons for the support which Filipinos have given to the United States in the wake of the destruction of the World Trade Center. In return, the United States, perhaps somewhat clumsily, has begun to help the Filipinos fight the Abu Sayef, which is composed mostly of brigands masquerading as Muslims.

Even though the Phillipines is no longer threatened by an insurgency such as the NPA that could take control of the country, all is not well. Since Marcos' ouster, bouts of political instability have held back growth. Many Filipinos still live in abject poverty, so that the raw material for future instability is still present. Still, the democracy movement sparked a rapid expansion of civil society, not only in Manila but elsewhere in which the environment has an important place.

Moreover, the industrial-strength loggers have moved on to Indonesia, leaving the Filpinos, armed with new legislation, in a better position to preserve what is left. Without much of a budget, they are making an effort, exemplified by the watershed above Cubi Point, which remains in a state of nature after the departure of the U.S. Navy. Some reforestation is also under way, though resources are scarce. The new and severe challenge lies in the reefs and fringing seas, discussed later in this chapter in a regional context.

The Rise of the Environment as a Security Concern in Southeast Asia

The Philippines, because of hard experience, became the first country in Southeast Asia to recognize the environment as an element of instability. That is not to say that the Philippine elites, upon whom policy and the economy depend, are convinced environmentalists, only that some of the elites now recognize a linkage between major environmental damage and instability. It is important to note, in this context, that the new president, Gloria Macapagal, has a track record of concern for the environment and its role in sustainable growth. How this concern will translate into policy remains to be seen.

In the rest of Southeast Asia, recognition was slow to arrive. In 1997, the annual Association of South East Asian Nations (ASEAN) conference in Kuala Lumpur, which brings together the security elite from think tanks and governments around the region, held its first ever panel discussion about environmental security. Sandwiched among well-attended panels, it drew an audience of nine people, while down the hall a standing room only crowd of nearly a hundred persons listened to a panel about piracy.

The panelists appeared to believe the environment did not pose much of a problem for the region, which was flourishing economically. Asked if the region was too complacent about the risks from environmental damage, a panelist suggested the questioner was out of touch with recent developments. In the past, such a criticism might have had validity, but no more. Matters were now in good shape. To support his optimism, the panelist cited a *joint declaration* issued the previous year by the ASEAN states, expressing support for conservation.

A few days later, the fires began. Set by men who apparently had not read the declaration, the fires would rage out of control in the forests of Borneo and Sumatra (and, for a time, in part of Malaysia) into the early part of 1998, when at last long-delayed rains finally penetrated the forest floor. The fires spread choking smoke (called the "haze" in the politically correct language of the times) across much of Southest Asia, closing airports, causing severe health problems for months, and contributing to Indonesia's mounting difficulties.

The fires, which were contemporaneous with the Asian financial crisis, prompted a new recognition among knowledgeable Southeast Asians that the environment might be a legitimate security issue. A year later, at a conference on the same subject, the panelists and their papers, presented to a large audience, struck a note of realism, often hard-edged.

In later visits, including one in the early part of this year, I have encountered widespread acceptance of environmental security as a valuable concept, relevant to the region's domestic and international concerns. The concerns are the familiar ones, the links with internal tensions and stakeholder conflicts and with external disputes, exemplifed respectively by Indonesia and the South China Sea, both discussed below. The belief that Southeast Asia had so far outstripped the rest of the developing world, that it had little in common with other regions, had given way to a recognition that some challenges, among them, the environment, resemble those elsewhere, particularly in Latin America.

1997 and All That—The "Young Dragons" of Southeast Asia

No country is more closely watched in Southeast Asia than Indonesia, which has periodically threatened to fall apart ever since the waning days of the Suharto regime, which was ousted from power in 1998 after a year of turmoil.

With a population of about 210 million people, Indonesia is the world's fourth most populous nation, after China, India, and the United States. About 90 percent Muslim, it is also the largest nation where Islam predominates. In the decade before the Suharto regime rotted away, Indonesia had become the fifth member of the elite class of the "young dragons" of Southeast Asia, along with Singapore, the Republic of Korea, Thailand, and Malaysia.

In this metaphor, the elder dragon was Japan, the first Asian economy to become a fully developed nation and eventually the world's second largest economy, a status which it still enjoys even though it has been mired in economic stagnation for a decade. Japan served as an inspiration for these Asian countries and, along with the United States, as a source of capital and as a destination for their exports, which became a motor for their economies. Each followed somewhat different economic models. Only the structure of the Korean economy much resembled that of Japan, although all of them, in varying degrees, invested in infrastructure as Japan had done on the way to its economic miracle.

The city-state of Singapore, the first "young dragon," subsequently became a fully developed country, with a per capita income above $20,000, the first country in the tropics to reach that status. Not long afterwards, the Republic of Korea entered the ranks of the "young dragons." By the early 1990s, Thailand and Malaysia had entered the lists, year after year enjoying, as Indonesia had just begun to do, high rates of real growth (defined as the percentage growth of gross domestic product [GDP], after subtracting the inflation rate) which often reached low double-digit figures. By 1997, on the eve of the Asian financial crisis and the beginning of Suharto's rapid slide toward oblivion, Indonesia was counted among the "young dragons."

To complete the ASEAN picture, in 1997 the small nation of Brunei on the north coast of the island of Borneo was rich, but only because it was awash in hydrocarbons in the manner of the Persian Gulf sheikdoms. The Philippines, still struggling with the aftershocks of Marcos's rule, remained an economic outrider though it was beginning to grow under the economic policies of Fidel Ramos, Corazon Aquino's successor. Vietnam, still captive of a Soviet-style economy and unwilling to imitate the economic policies of its huge neighbor and old enemy, China, lagged well behind. Laos, Cambodia, and Burma were led by a military narcocracy and remained hopelessly poor.

In 1997, the political picture was equally mixed. Japan was a democracy. Singapore had greatly prospered under a soft semiauthoritarian rule, of evident Confucian inspiration. Malaysia, like Indonesia, was predominantly Muslim (67 percent). It had effectively reduced old ethnic tensions with its large, more prosperous Chinese community, through careful balancing of, among other things, its electoral system and bureaucratic structures, though it was on the eve of important political tensions, involving generational change, far more than ethnic issues.

Three countries had recently become practicing democracies—the Philippines, Korea, and Thailand. In Korea and Thailand, democratization would soon serve as a buffer against political unrest when their economies plummeted in the wake of the 1997 Asian financial crisis. While the Philippine economy would suffer much less than those of its more wealthy neighbors, Singapore excepted, which would ride out the storm, bouyed by the reliability and general transparency of its banking and economic institutions.

In Vietnam, and even more so in Burma, a lack of democracy arguably held back their economies. Laos was so underdeveloped to begin with that it is difficult to ascribe a seminal role in its poverty to political causes and the Khmer

Rouge's genocide that so devasted Cambodia, that all other obstacles, among them the evident flaws in Cambodia's politics and political institutions, pale in comparison.

In gross terms, this brief survey reflects the Asian discussion, before the crisis and as it played out in 1997–98. (For that matter, it reflects then-prevalent Western attitudes.) The focus was on how to restore growth and, in the wake of the crisis, on institutional performance and, in the most troubled countries such as Indonesia, political structures.

Hobbled in varying degrees by what I have elsewhere called "the institutions of the day before," Southeast Asia looked for reforms, particularly in the financial sector. The aim was to restore, as soon as possible, *sustained* economic growth, and there was much discussion of transparency, capital flows and related matters, and hardly any about the role of the environment in development. In 1997, few people immediately recognized the link with *sustainability*, the ability of these economies to grow in the face of mounting environmental damage, which then remained largely the concern of a few environmentalists.

To be fair, the proper management of "brown" issues, pollution and waste, is a major part of squeaky clean Singapore's economic success, so much so that Singapore and Curitiba in Brazil have become *the* model cities for the developing world, for how to make large urban centers liveable and economically viable. Singaporeans will tell you, however, that except for nurturing green areas and botanical research, they paid little attention to "green" issues until they were enveloped in the smoke from Indonesia, the "brown" result of a "green" problem.

Among the "young tigers," Indonesia was the least fortunate. Korea and Thailand have rebounded, and Malaysia managed in the short term to insulate itself from the worst effects of the financial crisis through unorthodox economic policies. (None are out of the woods, if for no other reason than the global economic slowdown). In Indonesia, massive environmental damage, mostly man-made, is demonstrably contributing to the chaotic conditions and civil strife which have attended the nation's halting essay in a democratic transition from Suharto's rule.

Getting Down to Cases—Indonesia

The causes of Suharto's overthrow in 1998 are easy to describe, so much so that one wonders why it took so long to recognize that he was on the way out. They were the familiar devils of corruption and oppression, the Asian financial crisis, and, as the end neared, the evident exhaustion of the man and his personal vision of a restored Javanese empire. In Suharto's ouster, environmental change played only a minor role, but it did contribute to the rapid decline of his personal legitimacy in the last year of his rule.

In the main, the fires which ravaged Borneo and Sumatra were set at the instance of Suharto's "cronies," intimates who belonged to his inner circle of family and friends. Set to clear land for large palm oil plantations, the fires raged through the forests for months. Nothing could put them out except rain. But in the El Niño year, there was none, a stunning example of how man and nature may combine to produce disastrous environmental change. Indonesians blamed the Suharto regime for the fires, for the choking smoke, and for being powerless to control the fires.

Today, the Indonesian situation is vastly more complex, with the rise of separatist forces and communal and religious tensions long suppressed under Suharto. Conflicts have sprung up from one end of the realm to the other, in Aceh on the western tip of Sumatra, in Indonesia's great territories on Borneo and New Guinea, and far to the east in Sulawesi and the Moluccas. The evidence is persuasive, and this is what knowledgeable Indonesians are saying: that environmental damage is significantly contributing to the strife, to the point where it threatens not only Indonesia's economic future, but also the very integrity of the state.

Away from Java and Bali, which remain islands of relative stability, this is true in many places, even where the violence amounts to localized thuggery which local authorities either tolerate or cannot control. Where there is violence, usually it goes hand in hand with disputes among national and local stakeholders over resources, not just hydrocarbons and minerals, but over the exploitation of "green" assets, forests, fisheries, and reefs.

Last but not least, Indonesia's plight has international ramifications, because of its proximity and importance to smaller neighbors and East Asia in general. These ramifications are, in the first instance, economic. But they also include unwanted migration to Indonesia's neighbors, an increase in piracy in nearby waters including the Straits of Malacca, through which nearly half of the world's maritime trade passes, and a leadership vacuum in the ASEAN which had often depended for its cohesion on Indonesian leadership, now virtually absent.

The Kalimantan Test

From the air, on a clear day you can see Borneo forever, but from mid-1997 through early 1998, you could not see it at all because of the fires, which had left a choking pall of smoke over Borneo and much of Southeast Asia. Two-thirds of this huge island, one of the world's last great extensions of tropical forest, is Indonesian territory, though the north coast belongs to Malaysia except for the small principality of Brunei.

For some years, the Suharto regime had promoted immigration from the land-poor island of Madura to colonize Kalimantan (Indonesia's name for its part of Borneo) to work for logging, palm oil, and mining and petroleum interests. The original inhabitants, the Dyaks, had lived for millenia as nomads of the forest and riverbanks, fishing, hunting, and growing subsistence crops. Until the coming of the Maduese, the Dyak way of life had persisted fairly undisturbed, except for the introduction of Christianity and firearms, under Dutch and, later, Indonesian rule.

Working for business interests in Jakarta, the colonists made deeper and deeper inroads into the forest and along the rivers, pushing large numbers of Dyaks from their traditional habitat into shantytowns. Tensions began well before the first real clash in the summer of 1997, just about the time the fires began. The Indonesian press, skirting Suharto's censors, discreetly reported the fighting, suggesting that twenty or thirty people had died. Privately, knowledgeable Indonesians were saying that the death toll had reached at least 200 people, mostly Madurese. The Western press, with the exception of the *Wall Street Journal*, paid little attention.

Suharto's ouster in 1998 did not stop the conflict in Kalimantan, which had much to do with his grasping "cronies" (Indonesia gave rise to the phrase "crony capitalism") but little to do with Jakarta's politics. Indonesia's military, stretched thin by strife in regions such as Aceh, more intrinsic to Indonesia's survival, guards little more than the key places. In 2001, the official story still lags reality; it puts the death toll in Kalimantan at a few hundred. In Jakarta, one hears a much higher figure, put conservatively in thousands.

Reportedly, the Dyaks have had much the better of the fight. In the old days, they were headhunters. They have reportedly renewed the custom, infrequently by most accounts, as a form of exemplary punishment to get the Madurese to leave. Many thousands of colonists have fled to other islands. Some Dyaks talk of independence, but most appear to want something simpler—to be left pretty much alone in their forests.

Working with Indonesians, major international conservation organizations would like to persuade the government to declare most of Kalimantan a forest preserve. Strange as it may seem, making the Dyaks into the world's most feared forest rangers may be a reasonable solution for all concerned, the only sure way to preserve Indonesian sovereignty over this huge territory, with provisions to grandfather certain key economic activities. No one, for example, wants to close the oil fields (Dyak leaders talk of revenue sharing).

Kalimantan is a stark example of how environmental change can stimulate conflict, perhaps the only contemporary case where environmental change is the *principal cause* of an important conflict. So long as the nominally Christian Dyaks were more or less left alone in their forests, they coexisted with a quite different culture, that of largely Muslim Indonesia and its officials on the scene. Only after the Madurese colonists severely disrupted the Dyaks' habitat did serious conflict break out.

For the foreign affairs and national security community, Kalimantan is a wake-up call. Though the Kalimantan story is more striking than most, it is representative of a pattern evident in the developing world in which environmental change contributes to conflicts and to other tensions short of war and rebellion.

This leads us to what this book calls "The Kalimantan Test."

- In a given situation, does environmental change contribute significantly to violence or tensions within and among nations?

Aceh

Not all of Indonesia's conflicts meet this test. In Aceh, the environment does not appear to have played a significant causal role in a conflict dominated by historical, ethnic, and religious considerations. (Arguably, however, a reciprocal relationship has since grown up between the violence and the destruction of Aceh's forests and reefs.) Many of the four million inhabitants of Aceh, historically an independent sultanate, resent Java's dominance of postcolonial Indonesia. Seeing weakness in Jakarta, the insurgents fight for independence and, not incidentally, for control of the rich offshore gas fields and their revenues.

The insurgents espouse, among other things, a local Islamic extremism against the more secular (also largely Muslim) society in Java. Bloody clashes have taken place between Aceh's Muslim insurgents and minority Christian communities, similar to clashes on islands at the other end of Indonesia, notably in Sulawesi and the Moluccas. At this writing, in the first days of the campaign to rid Afghanistan of foreign terrorists and the local power structure which backs them, concern exists that violence in the name of religion, using Afghanistan as a rallying cry, will move inward from Indonesia's outlying provinces to Java, long thought immune to such strife.

Irian Jaya

In Irian Jaya, Indonesia's half of the immense island of New Guinea, the conflict meets the Kalimantan test, but the sources of violence are more complex, a mix of long-standing nationalism, habitat damage, and, lately, religion. Like the Dyaks, the tribes of Irian Jaya have lived in the forests and mountains since time immemorial and have not taken kindly to the destruction of their habitat. In contrast to the Dyaks, however, the numerous tribes of Irian Jaya never really accepted the rule of Indonesia.

The tribes, who are nominally Christian, have before them the example of the eastern half of the island, the neighbor nation of Papua New Guinea, granted independence without any struggle by Australia, the territory's former adminstrator. In these circumstances, neither the insurgents nor the Papuan Congress, a loose political grouping of the tribes, has shown much interest in autonomy and revenue sharing. For them, only independence will suffice.

Several months ago, the leader of a prominent international conservation organization told the author that the situations in Irian Jaya and Aceh, which boasts the last major stretch of tropical forest on Sumatra, were even more menancing and urgent than that of Kalimantan. He was speaking primarily of the unchecked environmental damage in both regions, but his remarks apply equally well to Indonesia's security challenge.

The tropical forests of Southeast Asia are disappearing, by some measures more rapidly than in any other region of the world, and nowhere more rapidly than in Indonesia. Journalists and environmental organizations have run out of adjectives to describe the destruction from logging, mining, riverine pollution, and the clearing of forest for palm oil plantations. The damage to reefs, fisheries, and marine life, discussed later in this chapter, is nearly as staggering.

In Indonesia, environmental security has become a two-way street. Damage of various sorts has stimulated violence, notably in Kalimantan and Irian Jaya, but also in localized situations where contention among stakeholders has contributed to lawlessness. But causation runs both ways: responsible Indonesian observers and the major international environmental oreganizations judge that the scale of conflict and lawlessness has stimulated a discernible acceleration in the pace of environmental damage.

Restoring Indonesia to a semblance of health is, in the first instance, a task for Indonesians in which they will need help from the international community. By now, it should be clear that policies which fix only on economic and security measures without addressing sustainability are no more likely to succeed in the future than they have in the past.

Surrounding Seas

The conventional wisdom holds that enviromental damage is for more likely to cause instability within nations than to engender international tensions. That is a most dubious proposition all the way around, as several chapters in the book suggests.

To be sure, open-ocean contention derived from environmental causes is hard to come by, nowithstanding the so-called "cod war" and tensions over whaling, including occasional intense encounters between nongovernment organization (NGO) vessels on "green patrol" and whaling ships from the few nations which still take a limited number of whales. Depending on how well or how badly international regulation of the takings of highly migratory species, such as tuna, turn out in the years ahead, we may see more such contention involving nation states in the future.

The focus of this chapter, however, is not the open ocean but what are described here as "surrounded seas," seas and gulfs which are in good measure surrounded by nation-states with intersecting maritime boundaries and exclusive economic zones (EEZs in the jargon). In the language of the Law of the Sea, these seas and gulfs are called "largely enclosed maritime areas." The Caribbean, discussed in another chapter, is a surrounded sea, and so is the Sea of Japan, where territorial disputes ensnarled fishing rights among Japan, Russia, and the Republic of Korea.

In the vast reaches of Southeast Asia, a number of "surrounded seas" exist. Here, we are concerned with the Gulf of Thailand, the small Gulf of Tonkin, and the very large South China Sea, which boasts notable parallels and a few key differences with the Caribbean. In each of them, environmental changes ranging from coastal pollution through reef destruction to serial overfishing, defined in this book as the progressive exhaustion of species and the largest fish (with the highest reproductive values), have become international issues, often with security implications.

The Gulf of Thailand

The Gulf of Thailand is virtually surrounded by four states. Here the territorial waters and the more extensive EEZs (which may extend 200 miles outward of the coast) are shared by four countries, Thailand, Malaysia, Vietnam, and Cambodia. Even at the mouth of the gulf, where it meets the South China Sea, the EEZs cover its waters.

In Hong Kong to present a comparison of environmental security risks in Southeast Asia and the Americas to the International Studies Association, I heard Professor Simon Gundy-Warr (National University of Singapore) present his definitive treatment of the Gulf of Thailand which analyzed not only the legal aspects, but also science and its real world implications for the headlong decline in the gulf's once flourishing fishery.

His insights led to revisions of an earlier draft, summarizing, in a vastly oversimplified way, environmental security in the Gulf of Thailand in a comparative framework. The four main points are:

- Even though there is no territorial dispute, the boundaries for territorial waters and the EEZ in the Gulf of Thailand have been settled in accordance with the principles of the Law of the Sea Treaty. The lack of an international fisheries regime for the gulf is a major factor in the serial depletion of the fishery, which has reached alarming proportions.
- Today each state is responsible for policing its share of the fishery. Thailand and Malaysia have a somewhat effective maritime patrolling capacity. So does Vietnam. Cambodia has none. Moreover, "hot pursuit" is not part of the package, so that a Thai patrol, for example, cannot act against illegal fishing if the boat escapes to waters under the responsibility of another state. And it is open season in Cambodia's waters.
- Stakeholder conflicts, particularly the contest between an endangered artisanal fishery and ravenous industrial fleets, mar whatever prospects remain for saving the fishery, even within the EEZ of Thailand, endowed with a significant patrol capacity.
- The Gulf of Thailand is an excellent illustration that settled maritime boundaries. While they take off the table of tension, one of its most common sources, it is not a substitute for international fisheries arrangements. To preserve or restore the fisheries in "surrounded seas" requires effective international arrangements for the security of the resource involving an agreed-upon sharing of sovereign responsibilities.

As anyone who has visited the region knows, pollution from agricultural, industrial, and sewage runoffs is extensive and spreads far out into the Gulf (the waters off the lower Kra peninsula—southern Thailand and Malaysia—are less affected). The Thais are making a serious effort, noticeably, for example, in the Chao Phraya River which flows through Bangkok, but there is a long way to go. A serious effort to restore the fishery, which is demonstrably affected by pollution, would require concerted action, perhaps with international help.

This pattern of serial overfishing and severe coastal pollution occurs in other "surrounded seas," such as the Caribbean and the South China Sea which boast huge threatened fisheries, though the destruction is less advanced than in the much smaller Gulf of Thailand. This pattern, discussed below, is a real, if often unrecognized, source of environmental security risk, particularly with respect to international tensions, short of war.

The South China Sea

The South China Sea touches the security of many nations. From the Straits of Malacca, the major sea lane from Europe runs through the South China Sea on its way to Taiwan, China, Japan, and the Republic of Korea. It is the scene of the most important maritime dispute on the planet involving China and its Southeast Asian neighbors, notably Vietnam and the Philippines, Malaysia and Brunei, and, under one interpretation of China's claims, a slice of Indonesia's EEZ.

China, it should be said at the outset, has stated that it respects freedom of navigation (including "innocent passage" of warships), nothwithstanding its claim on historical grounds, to almost the entirety of the South China Sea. In these circumstances, freedom of navigation at this time is not so much an active issue as a latent concern.

For authoritative treatments of the several interlocking disputes, readers should see the papers of Professor Lisselete Odgaard, Dr. Stein Tonneson, and Professor Ian Townsend-Gault presented to the Hong Kong International Studies Association Conference. Among other things, they reinforce the aptness of comparisons between the South China Sea and the Caribbean, even though the two regions present quite different geopolitical characteristics.

Journalism tends to treat the matter as a dispute over the Spratly island chain without reference to other key disputes, that over the Paracels, a separate group of islands, reefs, and shoals well to the north, and that over the waters of the Scarborough shoals, and similar offshore features found along the Philippine island chain (for example, Manila does not claim the Spratlys per se but rather the reefs and shoals paralleling the Philippines, such as Mischief Reef).

These disputes over the years have produced low intensity clashes, for example, China's military expulsion of the Vietnamese from several islands, Philippine action against structures built by the Chinese, and multiple seizures of fishing boats from one party or another, including Taiwan, which espouses a claim identical to that of China. Apparently none of the claims, though fiercely held, is conclusive from a legal perspective which ought to argue for negotiated or mediated solutions, the prospects for which are discussed by the cited authors.

The world focuses on the promise of large gas and oil deposits in the Spratlys and Paracels, though so far nothing has materialized except for Indonesia's Natuna gas field, a tip of which juts into the South China Sea well south of the Spratlys. The immediate risk lies in another quarter, in the exhaustion of the fishery, a renewable resource which has been a huge source of protein for China, Southeast Asia, and the international fishery.

In my view, the prospects for settlement of these disputes largely have been stymied by China's position, which has treated the territorial issue as legally settled in China's favor and hence not subject to negotiation. "There is no dispute," in the words of a youngish Chinese diplomat responding to a presentation about the dispute by a senior government official from a Southeast Asian nation. Legally, the diplomat indicated, the South China Sea is China's, period. (He did repeat Beijing's position that other matters, i.e., resources, could be discussed among the littoral states). Presumably the diplomat's response looked good on his record, though it did little that night for international comity.

Still, one avenue, that of "track two" talks (in the ASEAN idiom) of an informal nature among experts, unofficial and official, from the littoral states, has shown promise. Pessimists characterize China's participation as just for show, but perhaps that is not the case. China, after much initial reluctance, has engaged in talks about a "code of conduct" for the South China Sea among the concerned states, in line with China's position that the uses of the sea are a fit subject for discussion, independent of the "settled" territorial issue.

The code of conduct is intended by the ASEAN, which first proposed a draft in 1999 to reinforce international prohibitions against the use of force, but it may also involve the fishery because confidence-building measures, envisaged for any code of conduct, involve cooperative endeavors, which China reportedly is emphazing in these talks. Not yet an agreed-upon document, a code of conduct, if it ever became reality, might amount to no more than an expression of principles, useful, but without teeth to regulate the fishery.

Of course, Beijing might show enough self-confidence (and self-restraint) to change its position, as it has recently done with respect to the United States (in tandem with Washington, it must be said). Were China to agree with the ASEAN states to territorial negotiations, any solution emerging from the talks would still leave China with the largest share, though not with the roughly 80 percent of the sea which China's maps now show as its territory.

Are there grounds for optimism? I remain skeptical; constructive give-and-take with the United States, the other big guy on the block, is one thing, giving in to territorial concerns of smaller neighbors is another. There are more optimistic views, cogently set forth in Odgaard's paper. The recent joint Chinese/Vietnamese demarcation of the Gulf of Tonkin could be a harbinger for the Paracels and Spratlys, though the technical problems are far more difficult. And there is a story, sourced to well-placed scholarly circles in China, that Beijing will propose to the ASEAN a comprehensive solution of all the maritime disputes.

Trial balloon, or just rumor? Regardless, the fishery is in sharp decline, though statistics have been harder to come by than in the Caribbean. The industrial-strength fishing boats, if left unregulated much longer, will exhaust the resource, which is also threatened by cyanide fishing and dynamiting of the reefs, already weakened by a record warming of the tropical seas as a result of the last El Niño. This degradation, expressed in serial overfishing, coastal and maritime pollution, and reef distruction, has security implications. Stakeholders, local and national, will scramble for what is left of the resource before the money and the fish go away.

Notwithstanding a growing recognition of the risk reflected in statements about cooperation to conserve and restore the fishery, nations remain reluctant to enter into enforceable regional arrangements. There is no legal bar to them, even in the absence of delimited maritime and EEZ boundaries, as Townsend-Gault points out in his paper.

What holds governments back is the pyschology of soveignty. They are cautious about even modest adjustments in the traditional scope of sovereign activity through international regimes involving shared arrangements for policing across maritime boundaries and EEZs. In their absence, the fisheries will dwindle away with predictable costs for economies and societies. How, for example, can China and the ASEAN replace the protein? Imports are too costly; the handy answer, aside from going without, is to invest in more pigs and chickens, which bring other costs, notably their wastes and what they do to riverine and coastal ecosystems.

The Decline of the Reefs

Not far from the city of Cebu, a stretch of beach fronts a broad channel which separates the island from a neighboring island. A shallow reef drops off straight down into thousands of feet of water, pitch black in the late afternoon—enough to cause a diver to shudder involuntarily. In 1978, this reef, like others near to cities, was bleached, but most of the reef systems around the Philippines were still intact. Not so today.

With many more species of corals (sources give different numbers), the Pacific is a far more fertile place for coral than the Atlantic and the Caribbean. At a February 1998 environmental conference in Okinawa, the heart of the Ryukyu islands, which hosts the Pacific's northernmost large reef system, Japanese marine biologists shared their analyses of the risks to the reefs. The 1997–98 El Niño had just raised Pacific ocean temperatures to record levels, causing massive coral die-offs. In the Ryukyus, waters reached an unprecedented 30 degrees Celsius at the surface.

Some coral species tolerate very warm water; others do not. In the Ryukyus, El Niño caused a die-off of about 50 percent of the corals. Okinawa itself fared worse, with an estimated die-off of about 70 percent. With a local environmentalist, I toured what had been a flourishing inshore reef, bleached beyond recognition. Biologists would experiment with transplants, but as one said, transplanting coral is accomplished only "with difficulty." The best hope lay in the annual bloom of coral larvae, which carried on the currents, slowly regenerates the reefs.

East Asian and Southeast Asian reefs were severely affected. The May-June 2001 World Watch, drawing on authoritative studies, suggested that East and Southeast Asia reefs, harboring 30 percent of the world's coral reefs, lost about 34 percent of their coral, while the Coral Sea, south of Indonesia and north of Australia, lost only about 4 percent. By way of comparison, the Indian Ocean, with 24 percent of the world's reefs, lost a staggering 59 percent of its coral, while the Caribbean, with about 15 percent of the reefs, had lost about 22 percent from the spillover effect of El Niño.

So long as the frequency and intensity of the El Niños do not increase, and that is an unknown linked to global warming, over time the reefs can be expected to regenerate themselves, a phenomenon which is happening in a number of places. But this does not take into account the hand of man. The same article estimates that 11 percent of the world's reefs have been lost to the usual suspects, silting from deforestation, agricultural and industrial runoff, sewage, waste dumping, and fishing practices, e.g., industrial trawlers, cyanide, and dynamite fishing.

In the Philippines, an estimated million kilograms of cyanide have been injected into the reefs, a practice which kills corals as well as fish. In Indonesia, dynamite fishing, enabled by conditions of near lawlessness in many outlying regions, is causing an even more rapid decline in what is by far the greatest set of coral reefs under the jurisdiction of a single nation.

There is a curious dissonance in the way specialist publications characterize the risk. The UN's Global Environmental Overview (GEO-2000) makes but two brief references to the reefs, while international conservation organizations see their deterioration as a major challenge. For example, the World Wildlife Fund's "Global Two Hundred" lists no fewer than twenty-one major tropical reef systems of immediate concern, eight of them in the region discussed in this chapter.

In the real world, fish has been a major source of protein for littoral and especially the island states of Southeast Asia. Indonesia, according to World Watch, was accustomed to getting 60 percent of its animal protein from fish. In prosperous countries, consumers can still choose to pay higher prices, but in Indonesia and the Philippines, the depletion of the fisheries means going without. Artisanal fisheries, for generations a major source of food, employment, and revenue for the common man, have been endangered.

East of Bali beyond Wallace's line, where the flora and fauna of Southeast Asia give way to that of Australasia, exists the Wallacea, a region which includes Sulawesi, the Moluccas, and New Guinea. Named for Darwin's collaborator, Alfred Wallace, the Wallacea harbors what is said to be the earth's most prodigious marine habitat; even this remote sea is under seige.

Make no mistake, this is a security problem for Indonesia, less spectacular than Aceh but hard-edged nonetheless. Stakeholders of a sort, the reef bombers and poachers are only a step removed from piracy; reportedly some moonlight as pirates. Mostly unemployed youth led by local gangsters in search of the fast rupiah, they threaten not only the writ of the authorities, but also local communities which are trying to protect their artisanal fisheries or their tourist potential, largely unrealized east of Bali.

Earlier this year, in a visit to Jakarta, I spoke with a number of knowledgeable Indonesians, social scientists, NGO leaders, biologists, and conservationists. They did not hesitate in setting environmental security in an Indonesian context, describing how damage to the forest and the marine habitats is promoting violence, not just the big conflicts in places such as Kalimantan, but also among local stakeholders in the remoteness of Indonesia's 17,000 islands.

Indonesia, as noted earlier, is in the throes of a political transition toward an unknown destination. Authority in the outlying regions is one of the casualties. The Habibie government, the first of three post-Suharto administrations, faced a twofold crisis in the provinces, popular demand for greater local control (supported by an expanding universe of civil society institutions), coupled with a marked decrease in the resources available to the central government to administer these regions, resulting from the financial crisis.

In a law which took effect in early 2001, the government chose to massively decentralize but without transferring monies to provincial and local authorities. As a prominent NGO leader put it, "we want decentralization but what is happening is a 'devolution' of power to local authorities, with neither the money nor the training to carry out their new responsibilities." What he and others meant by "devolution" was not the Magna Carta but a wholesale dump of government functions to the periphery. Leaders of environmental organizations stressed that it was hard to convince local authorities that they now had responsibilities, once exercised by Jakarta, for conservation.

In the end, this may be a good thing; Indonesia is too large and too diverse for a highly centralized state in which Jakarta makes all the decisions. But in the current circumstances, Indonesian environmental groups and NGOs must scramble to build new alliances and partnerships with local authorities and communities. Money is part of the problem, but international conservation organizations are providing financial and technical support to local communities and Indonesian NGOs. The greater difficulty is a Hobbesian one, a general lack of authority and security away from the center, fueled by ethnic and religious conflicts and by an explosive increase in damage to the forest and marine environment.

Rivers of Contention: The Mekong

The giant Mekong River is the most important international river in Asia. Rising in the mountains of China's Yunnan province, it rushes through the valleys between Laos and Myanmar (Burma), enters Laos, then marks a long stretch of frontier between Thailand and Laos, ducks back into Laos, enters Cambodia, emptying through Vietnam into the South China Sea. Swollen by the monsoons, the Mekong and its tributaries (3,025 miles in total) water the great Southeast Asian rice basket which helps feed 60 million people.

The Mekong harbors all the major points of contention with respect to international rivers and the environment—the management of mountain watersheds which govern the quantity and quality of the river's water, plus the various "upstream-downstream" issues of water takings, among them dams of which China is fond, pollution, and silting. Consider also that the Mekong begins in China, a nation facing a water deficit in regions remote from the river which may amount to as much as 25 percent of China's demand for water by the year 2020.

A "water war," strictly speaking, is not in the Mekong's future. Which Southeast Asian state would go to war with their giant neighbor over water? But Southeast Asians have reason to be concerned that upstream dams, water takings, and other actions, will reduce the amount and quality of water available to them for flooding the rice fields and for municipal and industrial uses.

Upstream nations, especially powerful ones, have the upper hand. Southeast Asia is not comforted by Beijing's persistent refusal to join the four-nation Mekong River Commission established to monitor the river and coordinate its uses. The issue has a concrete dimension, the new Chinese dam at Daochaosan on the upper Mekong, the second in a series of eight dams, part of a grand design for hydropower and to make the river navigable to the sea, by levelling the rapids in China, Myanmar, and Laos.

China contends this will benefit downstream nations, by modulating the annual flooding cycle, which is precisely what worries its neighbors, who fear disastrous effects to agriculture and to freshwater fish species in Laos and Cambodia, which gets much of its animal protein from the Oreat Lake. The Mekong is provoking what this book describes as "tensions short of war."

The Tachin River Basin

Thailand is a nation which respects the arts of compromise. Mahidol University is in the process of "adopting" the Tachin River basin and watershed, an important river which harbors at the micro level (it is not "micro" for the stakeholders) most of the tensions that may arise within a single country over the uses of rivers and their watersheds.

Early this year in Bangkok, I spoke with a principal architect of this experiment. Under the rubric of environmental security (a phrase which has also found favor in Thailand), the university, with national and local encouragement, hopes to help (and study at the same time) with the resolution of stakeholder conflicts. The stakeholders include, among others, local authorities (who have taken on a more important role in the aftermath of democratization), local communities, indigenous groups, agricultural and industrial interests, NGOs of various sorts, and, in a manifestly Buddhist country, the temple authorities.

These festering stakeholder disputes center on water takings and pollution. Cleaning up important stretches of the river is a critical task. Though tempers have risen, no real violence is involved, but the future economic well-being of the region is at stake. It is too soon to talk of successes, but most of the stakeholders, who cannot resolve these conflicts by themselves, apparently see the intervention of a respected university as a sensible method of bringing people together to find a constructive approach to sustaining the viability of the river.

Policy Implications

This is a book of risk analysis, not a prescription for policy, about which the authors may reasonably disagree, though we are united in seeing environmental change as a nonconventional threat to security involving real people in real places. Still, if one accepts the premise that environmental change, sudden and cumulative, increasingly plays a role in stimulating violence and lesser tensions, within and among countries, these cases have implications for policy. For this reason, the concluding chapter summarizes some of the forward implications for American policy of these new, environmentally derived sources of conflict.

Taking into account the emergency into which mass murderers have thrust the United States, a few thoughts about Indonesia may be in order here. The world's fourth most populous country with the world's largest Muslim population, Indonesia has always been an important place, though its importance has gone largely unrecognized in the United States.

By the time this book is published, the world will be different, better than it is today, one hopes. But no matter how much headway America has made against today's terrorists, the survival of Indonesia and its generally tolerant style of Islam will remain a matter which directly touches our own well-being. Only the Indonesians can harness their problems, but they will need, for a time, the assistance of the international community and the United States in particular. Leaving things to the International Monetary Fund will not be a viable strategy.

Such assistance will be tailored to Indonesian realities, or it will be wasted. To restore the legitimacy of the state and provide for the security of its citizens, Indonesia will have to fend off extremists, protect minorities, and resolve issues of autonomy; make decentralization work, restore economic growth; and arrest the headlong destruction of environmental assets, a major contributor to insecurity and poverty. That is a tall order, but if Indonesia holds, as it has so far despite predictions to the contrary, targeted support for its reconstruction would not be a "giveaway" but an investment, not only in the future of this giant nation, but also in global security.

In this devil's brew of insecurity, conflicts over environmental assets are a prominent factor. The environment is joined at the waist in three issues, also crucial to restoring the legitimacy of the state: (1) sustainable development of local communities, (2) effective decentralization (or, in few places, workable autonomy regimes), and (3) the provision of acceptable levels of security away from the center. Free trade agreements or macroeconomic measures, important for the long haul, do not address these issues.

In the present emergency, the critical path for Indonesian/world community cooperation may be found in a concerted address to the cluster of issues enumerated above, which fit comfortably under the rubric of environmental security, at least as Indonesians describe the matter. Be clear that public-private partnerships, not old-style aid programs, are what is wanted. They would involve not only the Indonesian government and its international partners (the UN, World Bank, Japan, the United States, and others), but also civil society, beginning with the lively world of Indonesian NGOs, scientists, and local communities, and extending to respected international NGOs active in conservation, commmunity development, and rural health.

To this, it may be objected that it is too hard; we have a war to fight. Well, the West fought the Cold War with such tools. Sustainable development, at root, is a political strategy, designed to substitute hope for desperation.

Note

1. This chapter is the result of numerous interviews, discussions, and conferences in Southeast Asia and Japan from January 1997 to date.

CHAPTER THREE

Environment and Conflict in West Africa

Aondover A. Tarhule

During the past two decades, the view that environmental degradation and sociopolitical upheaval are interlinked gained significant support among a large section of policy, political, and security scholars. The concept expands on the classic, strictly military definitions of security, which focused on external military threats to the territorial integrity of the nation state[1] to include natural resources, human rights, and economic scarcities as potential sources of conflict. Thus, scarcity, mismanagement, and human-induced degradation of natural resources are seen as potentially capable of destabilizing a region or country to the same degree as an invading army or external aggressor.[2,3,4] In what follows, I examine this proposition in the context of West Africa, where the livelihoods of a majority of the people are intricately tied to the environment. With its myriad, environmental problems, West Africa offers insights into specific mechanisms and dynamics of the link between environment and security at several levels.

Two preambles are necessary before we can embark on this task. First, we must recognize that the premise, meaning, and significance of the link between environmental degradation and security are heavily contested.[1,5,6,7,8] Critics generally point to the absence of examples showing *direct* cause and effect, as opposed to prevailing apocalyptic predictions conditional on scenarios that may either never be realized or that may evolve differently. Without empirical evidence of a major war or conflict that can be attributed explicitly and largely to environmental degradation, the arguments for a security-environment link appears, at least to classical security scholars, rhetorical and evocative of a Malthusian specter, in which demographics are replaced by environmental degradation.[1,5] Some proponents of the concept (e.g., Percival and Homer-Dixon)[9] concede that such direct connections are rare because of the mediative role that physical, technological, economic, and social factors play in translating environmental degradation into conflict.[2] My approach in this chapter is to acknowledge that environmental degradation is *a* cause of conflict, not *the* cause of conflict.

Second, Barnett[7] makes the point that many security scholars are guilty of looking for environmental conflicts that have the potential to threaten the security interests of the developed world. The absence of such threats in their view undermines the validity of the environment-security connection. Essentially, this view places a higher premium on the security interests of the Western developed countries and disregards real threats elsewhere. For example, the fact that environmental degradation poses little threat to the United States, as Levy concludes, should in no way diminish the intellectual basis of the concept, nor its significance in other geographic contexts.

My premise is that the effects of ecological change cannot be neatly untangled from the sociopolitical and economic systems of which they are at once a part and product of. A sensible starting point for our analysis therefore is to consider social-economic and political systems (or intervening factors) as the coupler between environmental degradation and security. This link can be conceptualized in the context of systems theory. Classical systems theory consists of input variables that interact with and are acted upon by transfer functions to produce an output. In our analogy, environmental degradation is the input, which interacts with and is acted upon by intervening factors to produce a range of output, which might include security threats at national or regional scales, however security is defined. This is the framework I have adopted.

Finally, both "environmental degradation" and "security" mean different things to different people in different contexts. In what follows, I attempt to clarify the sense in which I use these terms in association with West Africa. Specifically, I pose and attempt to answer the following questions: (1) Environmental security for whom and from what? (2) What are the most likely sources of security threats and conflict in the region? (3) What specific mechanisms are most likely to trigger conflict at various scales? and (4) What are the dynamics and pathways through which the conflict may propagate? I then proceed to explain how and why environmental unsustainability may lead to conflict and security concerns and proffer a number of feasible proactive recommendations. This task is timely and important especially given the vulnerability of West Africa to climate change,[8] the recent (1999–2000) political debacle in Côte d'Ivoire, the ever expanding war among the littoral states, and Nigeria's teetering democratic experience.

Background on West Africa

The region known as West Africa consists of the sixteen member countries of the Economic Community of West Africa (ECOWAS). These are Benin, Burkina Faso, Cape Verde Islands, Chad, Côte D'Ivoire, Gambia, Guinea, Guinea-Bissau, Liberia, Mali, Mauritania, Niger, Nigeria, Senegal, Sierra Leone, and Togo. With the exception of four countries (Côte D'Ivoire, Ghana, Nigeria, and Senegal), all the other countries (75 percent) have occupied permanent seats on the United Nations' ranking of the Least Developed Countries (LDC), with Sierra Leone and Niger ranking at the bottom as the poorest countries in the world. Despite their common and utter penury, there are important differences in geography among West African nations. Rainfall diminishes rapidly from south to north with corresponding change in vegetation belts. Luxuriant tropical rainforests along the coasts give way to grass-dominated savannah landscapes that progressively diminish to the scantiness of the Great Sahara desert. Six countries (Burkina Faso, Chad, Mali, Mauritania, Niger, and Senegal) lie entirely or substantially within the arid to semi-arid Sudano-Sahelian/desert ecological zone. Major environmental problems in this zone include drought, land degradation (88 percent of soils are degraded, the highest percentage for any major region in the world), desertification, wind erosion, and dune invasion. The other countries of the region fall within the rain, humid/sub-humid ecological zone, where the major environmental concerns include deforestation fuel wood depletion, gulling, soil erosion, and mineral oil pollution in coastal areas.[10]

In terms of demography, population growth rate increased from about 2.7 percent per year between 1965–1980, to 3.1 percent since 1980. Hence, not only did the population grow, the growth rate itself grew! Total current population is about 220 million, with Nigeria accounting for about one-half of the population. By contrast, per capita food production in the Sudano-Sahelian nations declined by –3.3 percent between 1980 and 1997. The coastal states did not fare much better, growing at only 1.8 percent, substantially below the population growth rate. The gross domestic product (GDP) for the LDCs in the region was also negative for the corresponding period, decreasing at an annual rate of –0.5 percent. As a result, cereal importation and food aid rose at a rate of 3.9 percent and 7.0 percent, respectively. Even so, the estimated caloric intake is only 87 percent of requirements, the lowest of any major region in the world. At the same time, external debt as a ratio of GDP increased by 14 percent between 1985 and 1997.[11] Hence, by any statistical measure, West Africa is a poor region and continues to slide further into an abyss of poverty, mounting external debt, and declining food production.

Security for Whom and from What?

Quintessential definitions of security emphasize the threat posed by external forces. However, for all its chaos, postindependence West Africa has witnessed relatively few cross-border wars among its nations. This is quite remarkable, given the arbitrary manner in which colonial interests partitioned and constituted the states.[12] On the other hand, the rarity of interstate wars conforms to a global trend of decreasing frequency of interstate wars and increasing domestic wars.[5] A brief description of problem spots follows. Mali has had two armed conflicts with Burkina Faso over the Agacher Strip, a border region about one hundred by twelve miles in size, first in 1974–75, and again in 1985. In 1988, the International Court of Justice divided the territory apparently to the satisfaction of both parties.[13] In 1990, a border dispute between Guinea-Bissau and Senegal erupted into armed clashes. To the east, episodic clashes between Nigerian and Camerounian troops have been ongoing over the Bakassi Peninsula, and the dispute is presently in arbitration at the International Court of Justice at The Hague. Border disputes between Niger and Benin are also in arbitration in the world court.[14] Among the littoral states, Charles Taylor's perennial and ever-expanding war has generated serious tensions along the borders of Liberia, Sierra Leone, and Guinea. Other forms of cross-border military interference by foreign forces have been limited to covert support for local insurgents and rebel groups (also known as war by proxy, or by the more corporate euphemism, war outsourcing). For example, Charles Taylor's shenanigans in Sierra Leone have earned him U.N. sanctions, and Blaise Campaore's (Burkina Faso) not-so-covert support for Charles Taylor provoked ECOWAS rebuke and threats.[15]

In sharp contrast to the low impact, low intensity cross-border military maneuvers described above, internal wars and conflicts in West Africa have tended to be ferocious and brutal. For example, the Biafra secession bid in eastern Nigeria persisted for nearly three years (1967–1970), with more than one million people dead by the time federal forces bombed and starved Biafra into submission. Nearly 3,000 people were killed in the seven-year-long Tuareg Rebellion in Mali and Niger, which also generated more than 120,000 refugees.[16] An estimated 220,000 people perished during six years of fighting in Liberia.[17] The war in Sierra Leone characterized by horrific atrocities subscripted more than 120,000 child soldiers, displaced one-half of the country's 4.5 million population, claimed in excess of 75,000 lives, produced an estimated 100,000 mutilations,[18] and pushed the country to the bottom of the U.N. ranking of least developed countries. Rhoda Howard[19] provides insightful reasons for the greater ferocity of internal conflicts relative to interstate wars in Africa. Here, it is sufficient to note, as the above examples show, that internal conflicts are more common, likely to persist longer, tend to be more destructive, and therefore pose the greater threat to state security. Prudence compels me to mention the contrary view expressed by Herbst,[12] which hypothesizes that Africa is ripe for an outbreak of interstate wars that would pave the way for the emergence of new, realigned geopolitical units based on the evolutionary principle of survival of the fittest. However, as Joseph[20] pointed out, the scenario that Herbst describes is extremely unlikely because African leaders have enough problems governing their own already pluralistic or polarized societies without adding the problems of their neighbors. Somalia, for example, has existed without pretense of any apparatus of government since 1991, and none of its neighbors has made a play for it. Second, African nations are militarily and economically weak; few have the resources or the requisite political security and legitimacy to wage war. Third, international respect for the territorial integrity of nations periodically enforced by sanctions and military force, as in the case of Iraq/Kuwait, the Balkans, and Sierra Leone, discourages explicit military aggression among states. Ironically, the same code of conduct that urges respect for the sovereignty of nations provides a ruse for noninvolvement in conflicts that are construed to be the "internal affairs" of the affected countries. This rationalization contributed to international inertia and prevarications during the genocide in Rwanda, allowing an unprecedented slaughter orgy to rage unchecked. For the same reason, no attempt has been made to interfere in the crisis in the Sudan, which has been ongoing for nearly two decades. Even the Algerian crisis, which arguably resulted in the deaths of more foreigners than any other African crisis in recent decades, has attracted little international involvement. The message is ringing in its silence: the international community will not interfere as long as conflicts appear to be contained within the territorial limits of a country, especially when the strategic interest of major countries or neighbors are not threatened. Such is the sanctity (or hypocrisy) of the principle of noninterference. For West African nations and other weak states, the relevant point is that the greater security threat appears to be from within not from without. It goes without saying that this position does not entirely dismiss the threat of external aggression. However, it becomes self-evident from the foregoing that any meaningful definition of security threat for West African nations must focus on the factors and processes that give rise to internal conflict without entirely deemphasizing the threat posed by externally induced aggression.

The Environment as Source of Conflict

Homer-Dixon[2,9] hypothesized that various combinations of environmental pressures are likely to result in scarcity disputes between nations or between ethnic groups, civil strife, and insurgency because as resources become scarce or degraded, competition increases and may cross a critical threshold beyond which conflict is inevitable.

In this section, I attempt to show the mechanisms through which this plays out in West Africa, emphasizing the role that intervening factors play in translating competition for scarce resources into conflict. I approach this using four micro case studies: (1) drought; (2) competition for resources between sedentary farmers and herders; (3) agricultural practices and systems; and (4) environmental pollution.

Case I: Drought

As Africa's principal source of natural environmental disaster, drought has been responsible for more deaths, displacement, and human misery than all other natural or man-made environmental disasters combined. For example, as many as 250,000 people are estimated to have starved to death during the severe drought-induced famines of the early 1970s that affected the Sahel and Horn of Africa. Nearly 10 million faced starvation.[21] Between 1981–85, drought threatened 150 million Africans (about one-third of the population of the continent at the time), forcing 20 million to abandon their homes in search of food and water.[22] Of course, "drought" is not synonymous with "famine." Drought is a climatic phenomenon, while famine throughout history has been a problem of class, related to the organization of production and distribution.[23–27] However, in Africa, the difference is practically irrelevant, as the two appear to march in step. While the chilling statistics related to deaths and starvation are the most obvious and regrettable effects of drought, the phenomena evolve slowly, spawning effects that are multidimensional, multitiered, and diffused. Because of this complexity, the effects of drought on conflict do not involve a simple linear relationship that can be characterized with any certainty. Yet, to pretend that such connections do not exist is akin to proclaiming that objects that appear out of focus in a mirror do not exist! The following examples provide only a generalized tour of the various ways in which drought may contribute to conflict.

Drought played a major role in the Tuareg uprising in Mali and Niger. However, consistent with the systems theory framework, it must be mentioned that the causes of the Tuareg rebellion are deep rooted in history and politics. The light-skinned, fiercely independent, and proud Tuareg do not consider themselves black, or Negroid, a perception that is shared by the majority blacks who, in turn, treat Tuareg as "outsiders." Droughts occurring in 1973–74, and again in 1984–85, decimated the livestock of the Tuareg and severely imperiled the traditional pastoral lifestyle of the majority of the group. With their livelihoods wiped out, the Tuareg were forced into a life of destitution in shantytowns in Algeria, Libya, and Nigeria and in the capital cities of Niamey (Niger) and Bamako (Mali). This new way of life was completely alien to the Tuareg. Desperation and humiliation arising from their second-class status fueled the resistance to the central authorities of Mali and Niger—the two countries in which the majority of the Tuareg reside—and the demand for political autonomy. The crisis itself was sparked by the massacre of civilian Tuaregs at Tchin-Tabaraden, in Niger, in May/June 1990. The "incident" at Tchin-Tabaraden was merely a catalyst or pinprick that caused the floodgates of repressed sentiments and desperation to explode.

Drought also results directly in substantial percentage decline in the GDP, employment opportunities, and widespread sale of assets that may exacerbate income inequality.[28] The reason for this is that during a drought, poor people—usually rural inhabitants and landless peasants—are forced to sell their limited assets to the rich, creating a yawning gap between the two. After the drought ends, the rich, desirous to keep their new-found assets, usually set high prices that make it all but impossible for the poor to recover such assets.

The state suffers because agricultural failures reduce tax revenue while increasing expenditure on relief, social welfare, health and water supplies, and the logistical costs of food-related imports.[29] Security services are put under increased stress due to a rise in crime associated with temporary unemployment, migration, and increased destitution. The additional stress and expenditures force external and internal borrowing, and the adoption of structural adjustment programs with stringent conditions generally opposed by the majority of the people (Barnett makes a similar argument). For example, in 1980 the Senegalese government adopted a stabilization program in response to growing fiscal budgetary deficits due to the severe drought of 1979.[29] Opposition to such programs cuts across the strata of African societies. For many bureaucrats, stringent reform conditions that attend structural adjustment programs threaten to erode their power and privileged access to state resources. "To vulnerable groups—urban masses, fixed-wage earners, women, and students, [adjustment programs spell] inevitable hikes in the cost of social services to offset massive cuts in public sector spending."[30] As another example, famine, combined with the collapse in the international price of coffee, forced the government of Rwanda to adopt a structural adjustment program in 1994. The resulting hardships and decline in the standard of living have been implicated in the series of events that plagued the country during that period.[31] In both examples, we find the insidious but undeniable role of ecological stress, although such roles appear smudged in the final outcome (see also, Benson and Clay[29]).

Migrations induced by drought may pose the most serious threat to regional conflict and stability. Drought-induced migrations are intensifications of seasonal migration, which is a fundamental aspect of the West African economy and society. But there are several important differences. First, in normal years migration is voluntary, and the prime motivation is the search for economic betterment. This form of migration is demographically selective, involving mostly young males, and driven by rational economic choices. Itineraries are predictable and migrants painstakingly set the timing and select destinations to maximize their chances of securing meaningful economic engagement. These migrants make important contributions to the economy of the destination areas through the services they offer. Clear evidence for this appeared in Côte d'Ivoire during the presidential election debacle of 1999, when migrant workers from neighboring countries fled because of rising antiforeigner sentiments. These workers form the backbone of the labor force that gathers cocoa beans. Their departure therefore resulted in labor shortages that reduced the year's cocoa harvest from an estimated 1 million tons to less than 900,000 tons, with even more serious effects expected in the following year. Mr. Guapa Pekou, a hamlet chief, described the situation succinctly: "There is no one left to work the land . . . all of the Burkinabe have left. Look at the pods on these trees—but there is no one to pick them. Soon we will find we no longer produce the most cocoa in the world."[32] Clearly the significance of migrant labor extends beyond regions, to global economics.

Voluntary migrants usually return to their villages at the end of the labor season, taking with them wealth (the pay they received) to the source region. In this sense, migration is mutually beneficial to both source and destination regions and is an important component of regional economic interaction.[21,33] By contrast, drought-induced migration is forced, and the motivation is survival. During drought, the prevailing mentality is anywhere-but-here, and migration becomes haphazard. With no rational economic motive, migrants constitute drains on the economy of the destination regions because few of the destination communities are equipped to absorb large influxes of people. Furthermore, people displaced by environmental stress are less likely than voluntary migrants to return to their villages. This is because migration is usually the last resort in a chain of coping strategies. As the effects of drought and famine progress, a family will typically enter into loan agreements and then proceed to sell off their assets with land, the most valuable asset being the last. At this point the family has nothing left to offer and little to look forward to after the drought ends, except a life of serfdom. When such families eventually migrate, they have nothing to return to even when living conditions at the destination points become challenging.

Drought-induced migration weakens the fabric of social systems. Because able-bodied young males tend to form the majority of the migrants, rural areas are denied the benefit of young, energetic workers and the people most receptive to innovation and change. This process has resulted in a population structure dominated by women in the rural areas, who then have to shoulder even more responsibilities in addition to their already disproportionate share. Rural-urban drift therefore results in both the plundering of African agriculture and the feminization of poverty because rural areas are generally poorer than urban areas. Once they reach the cities, migrants swell the urban population, especially the ranks of the disgruntled and disenfranchised. The migrants contribute to the growth of squatter settlements, street begging,[36] unemployment, and high crime rates.[9,37] The rise in crime and added stress on inadequate city infrastructure eventually affect all city residents. Food shortages and prices escalate as the people who would otherwise be producing food in the countryside now wait to be fed in the urban areas. The urban population becomes more restive, forcing the government to adopt measures such as food subsidies or easing of restrictions on imported food and luxury items to placate them. These efforts tip the balance of payment against the government and divert resources from other programs, notably investment in rural infrastructure. Furthermore, importation of cheap but higher quality food kills the market for local food, further impoverishing rural farmers, while simultaneously cultivating a palate for foreign food that is all but impossible to break. Desperation and hopelessness leads to more rural-urban drift, and the downward spiral becomes entrenched and self-reinforcing. The state weakens from within.

It bears reiterating that these problems are the results of the general level of poverty in Africa, especially in the poor rural areas, and not uniquely associated with drought or even environmental degradation. Even so, the focus on drought is justified because, where is common, drought is the singular most important phenomenon in reducing the options and resilience of social and agro-ecological systems. Drought is therefore most likely to set in motion the chain of events described above.

Case II: Competition for Land between Livestock Herders and Farmers

Low literacy levels combined with lack of modernization of the agricultural sector have tethered economic and resource use activities to predictable, time-honored patterns developed through generations of experimentation and trial and error. Unfortunately, recent socioeconomic dynamics within the region have made some practices untenable, with serious implications for conflict. In this section, I first outline the traditional relationship between cattle herders and sedentary farmers—relying on the description provided by Van Apeldoorn[21]—then I identify external factors driving changes in the structure of this relationship; and finally, I draw out the conflict potential inherent in, and caused by, the changing nature of the relationship.

The economy of the Sahel ecological zone is dominated by pastoral migration, which is functionally adapted to the climatic rhythm. Herders drive their cattle southward during the long dry season in search of water and pasture and back northward during the wet season, spurred in part by the tsetse fly that infects the cattle with the trypanosomiasis bacteria (*Trypanosoma gambiense*), causing sleeping sickness. Herders pay little or no attention to national borders, as many are probably not even aware of the existence of such borders. Sometimes migration is local and involves movement from the interfluves to the riverine swamps or floodplains, known as *fadama* in Hausa (Nigeria) or *bas fond/wadi* in the francophone countries. Historically, herders and farmers related to each other in a pattern of interdependence and avoidance. During the rainy season, cattle have to be kept off cultivated lands. After the harvest, however, farmers are keen to have cattle kraaled on their lands, providing manure for the following year's cropping season. In return, the livestock feed on the crop residue.[21]

Serious conflicts and tensions have arisen between herders and sedentary farmers in recent years from a variety of sources. First, grazing routes have been taken over by farmers as a result of increasing rural population searching for more farmland. Cattle routes, fertilized by cow dung, proved especially attractive to farmers, who moved in and cultivated the land. But cultivation choked off the migratory routes of the pastoralists, denying them access to grazing lands for their herds. This conforms to Homer-Dixson's[2] relative scarcity hypothesis. Once established, farmers strongly oppose cattle traversing cultivated lands, while herders, faced with diminishing choices, resent this usurpation of their routes. After all, as far back their traditions go, they (herders) always roamed these lands on these same routes until the farmers interposed themselves along the path. In a Western context, a comparable analogue would be land developers who take advantage of cheap land values near airports to develop real estate and then turn around to complain about airport noise!

Second, restrictions to migration compel pastoralists to spend entire dry seasons in the drier Sahel, a practice to which neither beast, environment, nor herder is adapted.[38,39] In response, herders frequently and prematurely set fires to grasses in an attempt to stimulate regrowth of fresh grass for their cattle. Any crops still left in the field and not adequately protected by fire buffers are gutted in these blazes. These activities serve to deepen resentment against herders, so that the conflict becomes mutually reinforcing and self-perpetrating. Here again, it must be mentioned that the practice of setting fires to brush is endemic to the savannah, but it appears to assume special significance when herders are the culprits.

Third, competitions for riverine swamps constitute another source of conflict between herders and farmers. Prior to the onset of the Great Sahel Drought (1968–73), favorable rainfall conditions coupled with low population densities made shifting cultivation a feasible and pragmatic agricultural practice. At this time, riverine areas were generally avoided as farmlands throughout most of the Sahel because of the greater labor input required (except in Hausaland where the practice is centuries old). Consequently, herders grazed the floodplains unmolested throughout the dry season and came to recognize their preeminent access, if not right, to the floodplain resources during the dry season, irrespective of any tenures during other times of the year. However, beginning with the Great Sahel Drought, sharply decreased rainfall amounts and reliability combined with rapidly rising populations exposed the vulnerability of rain-fed agriculture and the food insecurity of the region, prompting the search for complementary agricultural systems. Faith that the Green Revolution would result in food self-sufficiency also proved misplaced, compelling policy makers to shift attention from the large-scale government irrigation projects to the so-called small-scale, farmer-managed irrigation using floodplain resources during the dry season.[40,41] Literally overnight, floodplain lands became prized territories and areas of intense agriculture to which herdsmen were no longer welcome, *especially* during the dry season.[42]

The combination of these factors has pitched pastoral nomads against virtually all the farming communities in the Sudano-Sahel ecological zone. Violent encounters from increasingly well-armed sides have become frequent, with the number of fatalities rising. In June 1999 Fulani herders, backed by Chadian rebels, invaded and killed over 400 farmers in the Taraba State of northeastern Nigeria. More than 300 other people were injured in these attacks reported by the the *Post Express*[43,44] and the *Guardian*,[45] July 3, 1999. In eastern Ghana, a one-year-old Fulani baby boy was shot dead on his mother's back while she received gun shot wounds in the back and neck, following a clash between Fulani herdsmen and some farmers that resulted from the destruction of farms by the cattle of Fulani immigrants. Three other people were shot in this attack.[46] In Western Mali, twelve people were killed in violent confrontation between herders from Mauritania and farmers in northern Mali over water use at a well.[47] In Niger, seven people were killed and forty-three injured in clashes between Fulani herders and Zarma farmers in the arrondissements of Tera and Birni N'Gaoure, near the village of Falmaye, ninety kilometers southeast of Niamey.[48] Other examples of conflict between herders and farmers are reported in Van Dijk[49] and Noorduyn and De Groot.[50] These events conform to the diffused, persistent, subnational conflicts hypothesized by Homer-Dixon and Rhoda Howard.[2,51,52,19]

Case III: Land Dispute

Thirty-seven prominent communal conflict areas have developed in Nigeria since 1990, and most remain unresolved.[53] Serious conflicts in terms of high casualties and the numbers of people affected have enmeshed the Zango-Kataf, Urhobo-Itsekiri, Itsekiri-Ijaw, Ijaw-Ilaje, Yoruba-Hausa/Fulani, Jukun-Kuteb, Tiv-Jukun, and Eggon-Tiv, which displaced 94,000 people in July 2001. Ethnic differences by themselves cannot explain the violence: some of the most intense fighting has been between people of the same ethnic group, including the Ibo-speaking Aguleri and Umuleri of Eastern Nigeria. In Western Nigeria, over 600 people have been killed in the persistent Ife-Modakeke crisis[54] that has at various times generated over 20,000 refugees. Idoma on Idoma violence in Agila and Tiv on Tiv violence in Toungo (both in Benue State) have also recorded fatalities in the several hundreds.[54] These conflicts prompted the BBC to report (Monday, June 7, 1999)[55] that Nigerian ethnic clashes were "out of control," echoing the sentiments expressed by the *Lonely Planet Travel Guide*,[56] which remarked "there is a real concern that the nation will burst into widespread chaos and violence at anytime." These proclamations are not alarmist, and the seriousness of the threats is reflected in the fact that by summer 2001, several Nigerian states were openly considering the formation of vigilante groups to help maintain peace.[57,58,59] There can be no clearer admission of state failure.

Competition for land is either a prime or close proximate cause in the majority of these conflicts. Elsewhere in Africa, the Hutu/Tutsi conflict, though incited by contemporary social relations, was essentially a conflict between "former landowners and former peasants."[19] Land is a particularly sensitive and contentious issue because it is an essential unit of production and sustenance. The cultural identity, economic status (or the potential of it), and political capital of whole tribes and clan lineages are determined by the size and type of their ownership of land. Among certain groups (e.g., the Tiv of Central Nigeria), land cannot be sold in the Western capitalist sense of the word because each generation is considered to be mere keepers (of the land) for the next generation. The guiding philosophy is that land is borrowed from future generations, not inherited from ancestors past. As a result of such spiritual reverence and economic ties, the loss of land from natural or human causes may translate to the loss of political and economic power, cultural identity, or a sentence to serfdom and bondage. Not surprisingly, no culture or family will willingly allow this to happen, hence land claims are often fierce and ferocious, driven by a sentimental, even irrational instinctive need to retain and preserve at all cost—in the most literal sense of the phrase.

The ten-year conflict between the predominantly farming, five-million-strong Tiv and the 200,000-strong Jukun in Central Nigeria present a good example of violent conflict directly attributable to the competition for farmland. Thousands of people have been slaughtered on both sides, and more than one hundred villages in the disputed area have been totally razed to the ground or reduced to rubble and now stand desolate and uninhabited.[45] Conflicts over land result from several factors, but perhaps the most important are land fragmentation and the traditional shifting cultivation system. Land fragmentation is the result of an intricate web of interaction between culture, low levels of agricultural technology, and land degradation. Referring to our systems theory framework, we find that here, environmental degradation is only one of several input variables that is acted upon by a set of intervening variables that I shall attempt to identify. The cultural variables relate to traditional inheritance laws that divide a man's land holding among his children (usually males). For example, a man who owns twenty hectares of land may divide this equally among his (say) four sons, each inheriting five hectares. In turn, each of the four sons would be required to divide their land among their own sons once the children come of age. Over time, the system produces successive generations of land holdings that are increasingly fragmented and reduced in size, forcing a shortening or complete elimination of the fallow period, with a concomitant deterioration in yield and economic returns.[60] Under ideal conditions, more intensive cultivation would be accompanied by greater input and more technology. But investments in methods of agricultural intensification are poor and inadequate. For example, worldwide, the number of farm tractors in use per 1,000 capita in 1998 was 4.45, compared with 0.21 for West Africa. The use of fertilizer—a key ingredient in modern agricultural systems—is about 9 kg/ha on overage in West Africa, compared to 262 kg in China and 69 kg in India.[28] In fact, in most countries fertilizer input has decreased, and agricultural extension services inherited from colonial administrations have collapsed, with corresponding adverse impact on yield. In the early 1970s, yields in the production of cereals in India and West Africa were approximately the same. Presently, however, the yields have increased by only 30 percent in West Africa, while in India they have more than doubled. As a result agricultural growth has been driven by expansion of cultivated area (about 0.7 percent per year), not through intensification of cultivation.[28] As an agricultural policy, expansion of cultivated area is inherently self-limiting because land, and in particular good farmland, is finite. Hence, the system produces two main problems. First, as the available supply of good farmland diminishes, poorer quality land—marginal land—hitherto used only for grazing, is brought into agricultural use. These lands are prone to environmental degradation including soil erosion, excessive leaching, and, consequently, low and unreliable yields. Second, as more land is cultivated, land available for grazing is reduced. Sooner or later, expansion brings farmers into conflict, either with herders as described above or with their neighbors. In either case, conflict results when farmers attempt to compensate for decreasing farm sizes or yields because every piece of land is subject to some claim to ownership. Outside urban centers, land ownership continues to be governed by vaguely articulated and unwritten traditional laws, which appear to rest fundamentally on occupation and use.[60] Historically, land surrounding a village was considered communal property, and power over its control and allocation was vested in the chief of that village. However, recent population pressures have forced a decollectivization of land ownership in preference for individual ownership. Foley et al., report that, in the floodplains of the Niger, renting land has become common, and litigation over family land ownership has begun to occur. Clearly, therefore, changes in the traditional ownership laws and transfer systems have not kept pace with the causative agro-ecological dynamics and pressures. Moreover, where changes have occurred, they did so without benefit of supporting statutory mechanisms and framework. In the event, contradictory and inconsistent decisions handed down by the law courts combined with long delays in court cases and have exhausted the patience of rural people and eroded what little confidence they had in the legal system. Violent confrontation results as villagers attempt to resolve the matter by their own means and on their own terms. In summary, land tenure and inheritance systems produce land fragmentation, which interacts with poor technology and improper farm management practices to produce land degradation, reduced yields, and conflict. In practice, the relationship may be circular rather than linear.

Another important though nonecological factor in connection with land disputes is that boundaries among various groups are notoriously porous and ill defined. In the absence of natural markers such as rivers and mountains, knowledge concerning the limits of one village or tribal land was usually retained by the elders and passed from one generation to the next through oral tradition. These unwritten records are highly malleable and frequently characterized by innocuously nuanced semantics, which over time become twisted to serve one or the other group.

Drought and deforestation also contribute to land degradation, reducing the total area of available productive land and intensifying competition. Deforestation decreases infiltration, induces surface runoff and soil erosion, and reduces the regenerative capability of the soil.[3,60,61] Complicating the problem, only small percentages of the land areas of West African nations are arable. For example, despite huge physical size, only 3 percent, 2 percent, and 0 percent, respectively, of Niger, Mali, and Mauritania are arable.[62] Hence, agricultural activities in these countries are concentrated in small tracts along their southern borders. During years of deficit rainfall, these already inadequate land areas shrink further still as desiccation renders even more land unsuitable for cultivation. At the same time, many farmers instinctively attempt to compensate for decreased yields by increasing ploughed land in marginal areas otherwise not suitable for cultivation.[63] This significantly increases their vulnerability, brings them into conflict with other groups, and deepens the sense of helplessness once the next drought hits. It is indeed a multidimensional complex web.

The significance of sociopolitical context and intervening factors is most clearly illustrated in these examples. To what extent can conflicts be considered the result of environmental change rather than the failure of the political system, given that a more responsive government and functional social system may have addressed these limitations before they metamorphosed into full-blown conflict? This question assumes special significance when one considers that conflict over land appears to be more prominent in Nigeria than elsewhere in West Africa. To what extent is the political system to blame? To take this argument to its logical conclusion, could good governance—whatever that means—have prevented all of the problems? The weight of present evidence suggests that the answer lies somewhere between the two positions.

Case IV: Mineral Oil Pollution

The Niger-Delta of Nigeria presents a good example of violent conflict induced by environmental degradation and pollution. Much has been written about this crisis, and excellent information on the background, causes, and evolution are contained in Osaghae,[64] the *Guardian* (UK),[65] Rowell,[66] Brooks,[67] and Hutchful.[68] The following account is summarized from Lee et al.[69] Minority ethnic groups in the Delta region, led by the 500,000-strong Ogoni, have long protested that not only has oil production failed to benefit them materially, but also that environmental degradation has destroyed the economic viability of the region. Environmental degradation in the Niger Delta results from gas flaring, above-ground pipe leakage, oil waste dumping, and oil spills. Prior to May 2000, approximately 75 percent of the gas produced annually was flared, causing considerable ecological damage to land/soil, water, vegetation, wildlife, and psychological stress and trauma for rural populations. Today, the Liquified Natural Gas Project has only slightly reduced the amount of gas flared. Gas flares produce soot, which is deposited on the roofs of buildings in villages in the vicinity. Rain subsequently washes off the soot, producing a cocktail of chemicals that infiltrates the soil, causing reduced plant yields and crop growth[68] and disrupting the wildlife. The runoff entering water systems kills off the fish. Without fertile land or water, indigenous groups lose their mode of survival and are threatened by food insecurity. Gas flaring causes psychological stress to villagers who, though having no electricity, struggle to sleep night after night under the brightly lit flares. In the context of larger environmental problems, gas flaring in the area emits some 35 million tons of carbon dioxide and 12 million tons of methane a year, making it a substantial contributor to global warming.

Gas piplelines, which should be buried, are instead laid above ground. Frequent and serious leakages from burst pipes kill vegetation and entire forests, rendering the land economically useless. Spills also pose severe fire risks. In October 1998, the world watched in stunned disbelief as several thousand people in the delta city of Warri were roasted when fire broke out after a pipeline was ruptured.[65] Oil spills and dumping of oil waste in waterways poison drinking water and kill fish and forests. Although Shell, the dominant player in the region, operates in more than one hundred countries around the world, forty percent of all oil spills recorded by the company are in Nigeria. During a ten-year period from 1982 to 1992, 1,626,000 gallons of oil spilled from Shell's Nigerian operations in twenty-seven separate incidents. A recently released report, by World Bank environment specialist David Moffat and Professor Olof Linden of Stockholm University,[70] shows that 2.3 billion cubic meters of oil from some 300 separate spills, almost one a day, pollute the delta.

Building canals and roads, largely to service the industry, has precipitated some of the most extensive environmental degradation. Roads block streams and floodplains, creating stagnant ponds of water, killing forests and flooding fields while simultaneously giving loggers better access to the area's fast-diminishing forests.

Economically, oil companies have created major distortions into the social and economic fabric of the local societies and perpetuated regional and class inequalities by creating lavish oil colonies in local areas where oil executives live, surrounded by abject poverty and environmental degradation (structural scarcity in Homer-Dixon's terminology). Local villagers lacking requisite skills to work in the oil industry are forced either to migrate to urban centers after being economically displaced or to work as low-skilled workers dependent on the oil companies. Household income in the area is below the national average, while health is substantially worse than in the rest of Southeast Nigeria.

The delta people have complained about the degradation of their lands since the early 1960s when a certain Isaac Adaka Boro, leading demands on behalf of the minority groups, was charged with treasonable felony and subsequently imprisoned.[71] The Ogonis became organized to protest the rape of their land in the early 1980s, but better so in 1990 under the aegis of the Movement for the Survival of Ogoni People (MOSOP) and the leadership of the erudite Ken Saro-Wiwa. From the beginning, the response of the federal government, which derives more than 80 percent of its foreign exchange earnings from oil, has been to suppress agitations and dissent with brute force. Ever present in the governments political calculus has been the fear that other ethnic groups would join the Ogonis in clamoring for greater autonomy and control over their natural resources. In that event, government ineptitude and insensitivity did more to ignite the present conflagration of ethnic conflicts than anything the Ogonis did or could have done. In November 1995, military dictator Sani Abacha hanged Ken Saro-Wiwa along with eight other Ogoni environmental activists, after the Ogonis presented their "bill of rights," in which they demanded oil royalties from the federal government for the period 1958 to 1993. World reaction to the hanging was swift and unequivocal; Nigeria was suspended from the Commonwealth, and both the European Union and the United States imposed limited military and aid sanctions. But the more serious effect was the worldwide publicity generated by the event. The thirty-million-strong indigenous population of the area rode the wave of public sympathy and organized (though not as one) several environmental and human rights movements to demand compensation and control over their resources. These efforts culminated in the "Kaiama declaration" of 1998, which got the full backing of prominent traditional leaders and politicians. Unfortunately, the groups frequently pitched themselves against one another (some claim due to the divide-and-rule tactics of Shell), leading to the orgy of killings throughout the Niger delta.

In 1999 alone, more than 200 people were killed in oil-related riots and in largely unreported clashes with the police and military; fifty incidents of kidnappings of Shell workers occurred within six months, and 150 Shell installations, depots, and pumping stations were occupied, closed down, or halted by protesting youths. All oil companies were banned from operating in Ijaw territory. State intelligence services in the region fear that the conflicts will continue to grow and have openly confessed that the safety of oil workers cannot be guaranteed.

With the return to democratic rule, the elected governors of the delta states—for the first time in the history of the federation—united in resolve to lend political muzzle, credibility, and organization to the demands for resource control. The tone of the demands has been uncompromising and chillingly eerie. For instance,

Governor Victor Attah of the Akwa Ibom state declared, "I will die to see that justice is done in this struggle and that equity is achieved. You cannot strip a man naked. If you don't treat him with respect, if you don't treat him equitably, if you don't give him justice, then you have turned him into a slave. And I was not born to this world a slave."[72] It remains to be seen whether such words are so much more political rhetoric—for which Nigerian politicians are famous—or a real threat. In either case, it would be irresponsible to ignore such rhetoric because they embolden and lend moral support to agitating youths. In a preemptory move, the federal government on February 7, 2001, challenged the legality of resource control by states in the Supreme Court. In the opinion of the federal government, all natural resources located within the territory of Nigeria belong to the federation and not to the state in which they are found.

It is uncertain whether even the Supreme Court decision will resolve the crisis. Officials of the Delta State government have begun making public statements that the struggle will continue irrespective of how the Supreme Court rules. According to Mr. Ighoyota Amori, commissioner for Education, an unfavorable Supreme Court ruling will "definitely not end the struggle, it will be the beginning, no matter how the ruling goes, we must be the winner. We are very sure we will win."[73] Such pronouncements forebode serious conflict ahead. In this micro case study, the progression from environmental degradation to violent conflict to political conflict is direct and unmistakable.

The case studies and conflicts presented are by no means exhaustive, nor are they the most important; they merely provided good examples of the salient connections I sought to illustrate. Some of the examples of conflict reported here are low impact, high threshold, or even benign in nature. However, they merit serious consideration because failure on the part of the authorities to control the situation may result in the conflict taking on a life of its own and becoming "irrational, cumulative, and manipulative."[50] As the Tuareg Rebellion in Mali and Niger revealed, it is politically suicidal and irresponsible to underestimate the significance of even benign but unidirectionally increasing tensions. Sooner or later, a critical threshold is crossed, resulting in serious conflict. Regrettably, such thresholds are usually recognized only in hindsight and are typically triggered by mundane events whose outcomes are disproportionate to the perturbing factors.

A second area of concern arises from the recognition of the role that deeply engrained ethnic vengeance and retribution justice can play in spreading an otherwise localized conflict. This was most powerfully and horrifically displayed during the war of attrition between the Hausa-Fulani and the Yoruba in Nigeria. As reported by the *Post Express*,[74] the crisis which began in the town of Shagamu, fifty kilometers north of the commercial city of Lagos in southern Nigeria, reenacted itself in Kano City nearly 1,000 kilometers further north as the Hausa/Fulani sought to avenge the losses suffered by their brethren in

Shagamu. As another example, exasperated by the government's inability to curb rebel attacks during the Tuareg rebellion in Niger, the Songhay militias of Gandakoy attacked and killed tens of Tuareg and Moor populations in reaction against the actions of banditry attributed to some elements of MFUA (Movements and Unified Fronts of the Azawad, a Tuareg rebel group). These examples show that conflicts in any region may ignite conflicts in areas that are not geographically contiguous. This leap-frogging effect may significantly shorten the time necessary for regional destabilization to occur. There is a risk that ordinary citizens may descend on law-abiding immigrants to avenge alleged mistreatment of their country folk, further eroding the fragile unity in the region. This risk increases with the degree of ethnic or national difference between the disputants, even if government forces do not become directly involved. Such diffused impacts and reactions blur the initial perturbing factor as other seemingly unconnected grievances get tacked on.

At this point, I anticipate the charge, "why focus specifically on environmental degradation given the apparent significance of the political context?" My response is that a general discussion of the whole motley of environmental, social-political, and economic problems is not helpful because important connections can become blurred in the complex web of interrelationships. Additionally, such an encompassing approach often has the unintended effect of creating a scenario that is intimidating, even intractable, especially when viewed alongside the resources available for tackling them. By focusing on selected micro cases in the preceding, I attempted to identify and illuminate salient features and dynamics of those connections between environmental degradation and conflict that have received little attention in the literature.

The Global Context

Our analysis would be incomplete if we did not make reference to how international relations factor into many of the processes described above. In today's global competitive environment, classic economic advantages like resource endowments and cheap labor have become subordinated to knowledge-based, dynamic, innovative production systems. Africa simply does not posses the requisite technical expertise to play at this game.[11] Progressive erosion of its competitive advantage has translated to loss of bargaining leverage in trade and multilateral agreements. As Africa moved to the periphery of the world economic order, deteriorating terms of trade combined with increased external debt burden have reduced the ability of many governments to meet their basic obligations to their citizens, resulting in the attrition of government, especially in rural areas. This attrition has, in turn, eroded government's legitimacy and control, creating a vacuum and permitting many groups and sectors to fall back into old feudalistic patterns. But if state structures have become too weak, insensitive, and/or incapable of meeting even basic obligations, should conflicts, which are, after all, legitimate means of effecting change, be encouraged as Herbst and others have argued? Herbst's argument is that many African countries are artificially supported in a comatose state with little hope of recovery. Such states, he reasoned, should be allowed to die naturally so that more viable and competitive states may emerge. I disagree with this position for two reasons. First, it assumes that the situation for many African nations is hopeless, but, as I shall show below, there are indeed many things that can be done at several scales. Second, conflict in Africa rarely tends to be driven by political ideology. Without an overaching ideology, local and transient interests may be accorded priority in ways that preclude the reemergence that Herbst envisions. I now discuss several steps that could be taken by various people to move beyond merely preserving the status quo.

Recommendations

There is no shortage of recommendations to Africa's myriad problems. The World Bank Technical Reports Series contains some of the most innovative and pragmatic recommendations to problems such as deforestation,[60] agricultural development,[28] water resources,[75] pastoralism,[38] and the impact of drought on sub-Saharan African economies.[29] It is difficult to match either the level of expertise, the diversity of perspectives, the depth of insight, or the micro/macroeconomic fundamentals contained in those reports. In addition, numerous reports and recommendations from specialist research agencies and organizations such as the Food and Agricultural Organization (FAO); the Comite Inter-Etats de lutte contre la Secheresse dans le Sahel (CILLS); the Center for Agricultural, Hydrological, and Meteorological Applications (ACRHYMET); the African Center for Meteorological Applications for Development (ACMAD); African Development Bank, government commissioned studies, universities, nongovernmental organizations (NGOs), and individual and freelance researchers address various aspects of Africa's environmental problems. But alas, there is no user manual for translating aspirations into practice. In what follows, therefore, I elected to focus only on those factors that have received inadequate attention in the literature. I approach this from the perspective that various stakeholders can address different aspects of environmental problems at various levels.

Vulnerable Groups

Vulnerable groups—especially women and rural dwellers—must organize themselves into cooperative associations to increase their visibility (and therefore political clout) and also benefit from shared experience and networking. Anybody familiar with the pattern of rural settlements in Africa will concede that the fragmentary and dispersed nature of such settlements makes it extremely difficult to provide the rural areas with facilities such as water, even if the political will and economic resources for doing so existed. By organizing and deciding on their priorities as they see them, rural dwellers will:

- take the proactive step in demanding for services, rather than docilely waiting for help;
- the cooperatives will constitute conduits of information for the dissemination of the results of successful and effective demonstration projects, weather forecasts/information and new technologies; and will
- have an additional forum for conflict mediation and resolution.

NGOs can play important roles in helping the rural areas to organize.

NGOs

NGOs make important contributions to the socioeconomic development of West Africa. These organizations are uniquely placed to introduce changes in ways that governments may not be able to do, for political reasons. First, they tend to work at the grassroot level and so come to know the people, the culture, and problems intimately. Second, the better NGOs are, and are perceived to be, politically neutral. This trust and perception can be utilized to encourage reforms in cultural practices and belief systems that would otherwise not be possible. More specifically, NGOs can:

- help rural inhabitants and other vulnerable groups to organize into cooperative associations;
- provide information enlightment campaigns in rural areas and technological resources (e.g., traveling cinema shows on new farming methods) that would not otherwise be available. These efforts will need to be coordinated with the relevant government agencies to avoid allegations of inciting the people;
- institutionalize vocational training in the rural areas. Many NGOs presently offer some form of vocational training in areas of interest to their project (e.g., pot making, stove construction). However, this has tended to be ad hoc, lacking coherence, continuity, and defined purpose. There is room for organization devoted solely to teaching vocational skills as a means of reducing dependence on natural resources;
- develop and maintain a database of conflict; and

- serve as conduits for international donor and funding agencies to channel money and resources directly to rural communities rather than the present government-centric policy that is riddled with corruption. Such funds have in fact served to perpetuate the very vices that they were meant to change.

International Donor and Funding Agencies

It is no secret that several African countries run almost entirely on external funding. International funding agencies are therefore in a unique position to influence changes in these countries. Unfortunately, to date, these agencies have a mixed record because of the existing modus operandi. For example, by arbitrarily setting the criteria and priorities for research and programs, international agencies sometimes impose solutions on rural communities. Countries have been forced to accept aid that they don't need because such aid is used to relieve the donor countries or agencies of excess inventory, or to capture future markets.[76] Having said that, many aid agencies are genuinely committed to help. These agencies and country may help by:

- funding grassroot cooperative associations directly rather than insisting on channeling funds through corrupt governments;
- funding basic applied research and research on appropriate technology (e.g., drought monitoring, information dissemination, education, and farm technology);
- funding the development and maintenance of a database of conflict in the West African subregion as a resource for scientific research and development planning. Political scientists and security scholars can browse, query, or study such a database to reveal causes, trends patterns, and connections that may otherwise not be obvious. Research based upon such empirical data would be more comprehensive and persuasive than those based on isolated case studies, or worse, back-of-the-envelope estimates. Politicians and policy makers can utilize the results of such studies to weigh the cost and benefit of various courses of action versus the consequences of inaction. To be free of bias—or the perception of bias—and political influence, the database would be most useful if it were maintained and funded by an independent and autonomous organization rather than a government agency;
- withholding funding for corrupt and ineffective governments. Such funds merely serve to prop up and prolong the government. The rapidity with which several governments fell once funding from the Cold War era ceased is proof of the efficacy of this approach;

- providing expertise on conflict mediation and resolution and disaster preparedness;
- providing expertise on land use reform, land tenure, and modern agricultural practices; and
- assisting progressive countries in debt relief through arrangements that ensure investment in social and environmental needs.

African Governments

In the end, African governments have to be responsible for their citizens. None of the recommendations above will succeed unless governments create the necessary enabling environments. Governments must increase their presence in rural areas, through programs that result in positive impacts on the day-to-day lives of people. Specific recommendations related to the environmentally induced conflicts include:

- developing security networks that involve community participation. Even at their most effective and functional form, the security apparatus of most West African nations will not extend to all geographic areas, especially rural areas. There is a need therefore to develop security networks that build upon customary forms of peace maintenance and greater interaction between formal security forces and the grassroot community-based organizations. A number of scholars may understandably squirm, interpreting this suggestion as advocating a police state. This is not true. Toward the end of British rule in Nigeria, the colonial administration maintained a staff of only 700, having effectively devolved authority to traditional rulers, without creating a police state. The essential idea is to avoid the power vacuum created by the attrition of government, by providing recourse to arbitration or opportunities for some semblance of judicial redress;
- recognizing the environmental causes of conflict. Denial is a disturbing but all too familiar characteristic of African politics. Insecure and disconnected from the people, some African governments prefer to view public discontent as threats to their governments orchestrated by political rivals, rather than examining the substance of the complaints. This policy informed the actions of successive Nigerian governments in the complaints by the Niger Delta people. Despite decades of environmental abuse, it took the hanging of Ken Saro-Wiwa and international pressure for the government to acknowledge the existence of the problem. On July 19, 2001, President Olusegun Obasanjo of Nigeria is reported to have blamed the introduction of Sharia in northern

Nigeria, the crisis in the Niger Delta, and the upsurge in militant ethnic organizations to "well orchestrated plots by unnamed sections of the country to secede,"[77] effectively denying the existence of any substantive basis for discontent. Similar sentiments were expressed by Balarabe Musa, a former governor, who described "organized violence" as "ploys to destabilize the country [Nigeria]"[78] and by governors of the restive western region who concluded that "there is an attempt to foment some problem by those we believe are fifth columnists, who we believe have infiltrated the nascent democracy we are in."[79] In Niger, the government insists on conducting a national census, which requires people to remain in their villages, even though droughts and famine have forced hundreds of thousands of people to flee their villages, and even the country and the government to appeal for international food aid;[80,81]

- implicit in these denials is the arrogant view that the government is infallible and what must be the political equivalent of Newton's first law of motion (read upside down) that the people are content (no matter what) and would remain so unless instigated by political rivals! The truth is that like the environment, individuals and societies can endure abuse and intense competition only up to certain thresholds. Denial—and therefore prompt and appropriate response—allows discontent to mount and fester until it erupts into violent conflict. This simplistic interpretation explains conflicts without the need to invoke the clandestine activities of mysterious political rivals;
- holding multinational corporations to internationally accepted standards;
- acknowledgement is a necessary first step that must be accompanied by strategies to address the source of conflict. In the context of environmental conflict, common sources include natural disasters such as droughts, competition for scarce resources (e.g., land, water, and forest resources), and environmental degradation. The solutions to these problems will require cooperation from all stakeholders.

Drought

It is imperative that Africa as a whole, and the Sahel in particular, increase its capacity for dealing with drought. Toward this goal, Van Apeldoorn's position is succinct: "with any disaster, the magnitude of damage or loss incurred by a society is a function of that society's level of knowledge about the disaster, as well as its degree of preparedness and use of scientific information. The capacity for dealing with droughts depends, among other things, on the extent to which: (1) the problem is understood; (2) such knowledge is accessible to potential victims and policy makers; and (3) that understanding is put into practice."[21]

Much remains to be revealed concerning the causes and physical mechanisms responsible for drought, but progress has been made to the point where medium- to long-lead climate forecasts of African climate are now possible, with acceptable skill. Even more information is available concerning the multifarious impacts of drought on socioeconomic and ecological systems. Regrettably, the potential social and economic applications of the results of drought research have lagged behind the scientific achievements. Two problems in particular remain poorly addressed: availability and accessibility of the results of drought research to end users and vulnerable groups and the capacity of the end users to act upon such results. Of 600 farmers in Nigeria, Niger, Burkina Faso, and Mali interviewed by Tarhule and Lamb (forthcoming), less than 10 percent had heard of, or utilized the result of, research in their farming activities. O'Brien et. al., reported similar results for the Southern African Development Community (SADC).[82]

For vulnerability to drought to reduce, dissemination strategies that permit the results of research to permeate down to the people, enabling them to adopt precautionary action, must be articulated. Research is needed that focuses on appropriate knowledge transfer mechanisms, taking into account the cultural beliefs and practices prevalent in West Africa. Indeed, there has been some progress in this regard. The New Radio and Internet Technology for Communication of Weather and Climate Information to Rural Communities for Sustainable Development in Africa (RANET) is a five-year project funded by USAID. The project focuses on (1) providing Internet connectivity in rural areas to enhance the acquisition, exchange, sharing, and dissemination of critical meteorological and hydrological data within African nations; (2) organizing training workshops to increase the use of such data in food production, water resources management, public protection, health, and policy making; and (3) providing cheap windup radios to rural communities to facilitate their receipt and use of meteorological/hydrological information. This is a step in the right direction and similar efforts need to be initiated and supported. The point is that African governments, along with donors, funding agencies, and researchers, must develop a network of information dissemination, not only as it relates to droughts, but also other environmental hazards such as desertification, deforestations, soil erosion, and efficient farming strategies.

Information alone is not enough and must be combined by appropriate response strategies. It must be acknowledged and appreciated that traditional methods for dealing with environmental problems developed from generations of trial and error and are therefore not flexible enough to respond to new and rapidly emerging sets of problems. Unless farmers have the capacity to take action based upon the information provided, such information will remain underutilized. For SADC nations, CICERO[82] concluded, "while dissemination efforts have clearly been inadequate, they do not appear to be as consequential as prob-

lems related to the capacity of small-scale farmers to respond to the [climate] forecasts." Constraints to the capacity to respond to drought lie in economic and social structures rather than uniquely in a lack of information. Access to credit, seeds, fertilizers, draft power, transportation facilities, and markets shape the ability of farmers to respond to climate information.

Moreover, because drought is integral to the climate of West Africa, drought planning and drought preparation must also be ongoing activities. Even the most efficient fire service will prove inadequate if fire victims are unfamiliar with evacuation procedures and exit routes. International donor agencies may be receptive to these activities if it is demonstrated that they are less expensive in the long run than emergency relief efforts.

Urban Areas Can Be Improved Only When Rural Areas Become Habitable

African governments have proved astonishingly insensitive to the needs of rural areas and inattentive to the significance of the linkage between rural and urban economies. In any event, they continue to perpetuate and reinforce urban bias in economic and planning policy for political expediency at the expense of sound urban-rural integration for regional development. The result has been the unprecedented emptying of the rural areas as people migrate to the urban areas. Burgeoning urban populations have commanded even more political clout demanding greater attention. As long as response strategies weaken rural areas in favor of urban areas, migrations will continue. Paradoxically, therefore, urban quality and status can only improve when rural areas become habitable (or totally emptied!).

Unfortunately, "improving the rural areas" is a vague and all-encompassing phrase that is useless as a practical policy, especially considering that these areas do not have even the rudiments of a modern infrastructure. Scholars have consistently advocated better network connectivity, water supply, sanitation health facilities, and functional markets. The scarcity of resources for dealing with the problems necessitates prioritization, which ideally should be informed by the fact that improvements in targeted key sectors with multiplier potential may spur developments in other sectors. Education, for example, is critical because an enlightened citizenry will be better able to understand and use the results of scientific research, and they will be more receptive to innovation and change, including gender-sensitive practices. The *type* of education is key. African governments have made the mistake, inherited from the colonial experience, of adopting a parochial view of education as preparation for white-collar jobs, to the neglect of vocational training. But vocational education is precisely what Africa needs, given the limitations of opportunities in white-collar professions and the concomitant significance of the informal sector. Drive through any city in Africa, and you notice the proliferation of roadside mechanics, electricians, technicians, and even health care workers, most of whom received no formal education. These dedicated workers perform essential services, but they also cause grave harm on occasion due to the lack of training. Governments, communities, and NGOs have important roles to play in establishing vocational institutions and encouraging their acceptability, especially in the rural areas. Vocational institutions have the potential to provide skills that lead to self-employment or improve opportunities for gainful employment besides or in addition to farming. The goal would be to reduce vulnerability to, and dependence on, environmental resources.

Limits of the Agricultural Sector

Rural development planners and policy makers must be especially aware of the fact that there is a limit to the capacity of the agricultural sector, in its current traditional form, to absorb new farmers. For decades, African agriculture has been characterized by the trinity of land, labor, and the belief that it will rain, and by maintaining ecological balance through shifting cultivation or pastoral nomadism. These strategies are no longer sufficient or even an acceptable basis for agricultural planning. Labor input measured solely in terms of the numbers of bodies without commensurate increase in the technological and modern input and the industrial capacity to process agricultural produce is unlikely to significantly increase productivity. In fact, because other critical factors upon which agriculture depends (e.g., land and water) are finite, continuous increase in labor input will eventually cross a saturation point beyond which marginal returns become negative. This threshold has already been crossed in many parts of West Africa, and additional numbers of farmers can only lead to conflict by intensifying competition for land and other resources. Efforts must be geared toward increasing agricultural intensification and yields while simultaneously reducing the number of farmers and increasing industrial capacity to utilize agricultural produce.

Local agriculture must be made viable by attracting private investment in modern technology and reducing dependence on imported foods, especially those for which local substitutes exist. For example, according to ECA statistics, Africa's wheat import rose steadily at 5.6 percent per year in terms of quantity and 8.1 percent in terms of volume. Most of the wheat is imported for the production of bread, a commodity that was alien to most of Africa at the beginning of the twentieth century. In any case, both the technology and expertise for producing bread from substitute flours exist, but African entrepreneurship has found it easier and more profitable to telex suppliers in the United States and Europe than to mobilize supplies from fragmented and complex local production systems. The effect of such massive importation has been devastating for both balance of payments and the local agriculture. Indeed as several studies[83,84] show, food aid and food importation promote the development of underdevelopment. In an uncompromising tone, ECA concluded, "by signing a food aid contract, [Third World] countries have signed a contract to support the American agricultural sector and to disorganize their own agricultural systems." For example, "Niger, which recorded a shortfall of CFAF 6,300 million (US$18 million) in 1986 would have shown a surplus of CFAF 1,700 million (US$4.8 million), had the country not imported foreign cereals, particularly wheat, for CFAF 8,000 million (US$22.8 million)." Exasperation at this trend compelled Gunilla and Beckman[84] to exclaim, "If Africans want to eat bread, let it be! But they must eat bread made from their own locally produced raw materials." This should be the overall sentiment governing Africa's food policy.

Land Tenure, Land Ownership, Accessibility, and Transfer Systems

Because the economies of West African countries are dependent on natural resources, it is logical that competition for those resources should occasionally result in conflict. Different stakeholders have different interests in the use of resources, and not all of these interests can be fulfilled all the time. As a result, adequate institutional arrangements are required to determine rights and access to rural resources, such as land, water, trees, and wildlife, as a prerequisite to agricultural development and food security.[85] Radical land and resource use reform projects are inherently controversial and politically risky and need to be executed in a comprehensive framework. For reform projects to succeed, the necessary institutional and social-political environments have to be favorable and appropriate adjustments made in the macroeconomic environment. For example, policies that directly curtail or force farmers to abandon shifting cultivation must be preceded by massive public education and the availability of inputs for intensive agriculture. Similarly, adequate arrangements must be made to accommodate grazers that may be displaced as a result of policies that affect pastoral herders. The ideal situation would be to make the new proposals sufficiently attractive that the affected groups demand them or market forces drive them. For example, nomadic livestock herding in its present form and practice will continue to bring herders in conflict with farmers. It appears that barring cultural inertia, a profitable business opportunity exists in selling fodder from more humid regions to herders in the Sudano-Sahel region, nullifying the need for seasonal migration with cattle. Successful adaptation of such novel but alien ideas will require political dexterity and skills that may challenge the capacities of many governments. Fortunately, the Sustainable Development Department of the FAO presently assists member nations in restructuring their rural institutions and in the identification and preparation of projects on land tenure institution building for food security and sustainable rural development. Support has been provided in developing and adopting policies, programs, best practices, and tools for land tenure to facilitate access to land, ensure user rights, and improve land markets. West African governments must avail themselves of such opportunities and expertise.

Food Security or Food Self-Sufficiency?

African intellectuals and policy makers need to question technical concepts and goals to ensure that they are relevant to their particular situation. The debate between food security and food self-sufficiency provides an example. The United Nations Economic Commission for Africa defines food security as access by all people at all times to enough food in quantity and quality for an active and healthy life. Its essential elements are the availability of food and the ability to acquire it.[63] On the other hand, food self-sufficiency is interpreted as self-reliance or the ability to produce adequate food nationally or regionally to satisfy basic needs at all levels. The implications of adopting one policy over the other are reasonably clear: if the goal is food security, African countries will have to produce enough cash crops for export and use the export earnings to import food. If, on the other hand, the goal is self-sufficiency, agricultural productivity will need to increase to meet basic needs. Some scholars[26] argue that food self-sufficiency is not a useful objective because it does not admit that in some cases food security can best be achieved by some combination of food imports and expanded agricultural production. This argument has some merit in the context of West Africa because despite huge territories, only relatively small proportions of land areas are conducive to agricultural productivity, especially in its present inefficient form. Increased technical expertise and technology will increase cultivable area but only to a point, beyond which serious environmental degradation results, with consequences for conflict. Food security presents the difficulty that African economies are unable to generate enough income to permit their households to buy food either from their own countries or from other African countries. Which goal is adopted must depend on each country's realistic analysis of its capacities and the relative risks involved with both options.

Conclusion

Internal conflict, rather than external aggression, poses the greater threat to the security of West African nations because of the pluralistic nature of their societies, deeply ingrained historical ethnic mistrust, and the high dependence on environmental resources. In this part of the world, deterioration or scarcity of the environmental resources upon which sustenance so tenuously hangs spells the difference between survival or mass migration, a life of despondency or death. This dependence explains the ferocity of the competition for scarce environmental resources. The effective and sustainable management of environmental resources is therefore key to improving the quality of life and mitigating conflict.

But environmental conflicts as well their implications on society originate and thrive within specific social-political and economic environments. In that sense, environmental degradation due to natural or anthropogenic change contributes to, rather than causes, change. However they originate, conflicts,

> undermine the promise of African democracy and development, contribute to political decay, and facilitate state collapse. As conflicts drag on and escalate, they can cause widespread violations of human rights and exacerbate famine conditions . . . wars divert scarce resources away from social services, disrupt trade, discourage tourism, and contribute to the breakdown of family structures. The pervasiveness and persistence of conflict also have grave psychological consequences as children are traumatized or become accustomed to a culture of violence.[85]

West Africa and, indeed, SSA must develop effective ways of preventing conflict. Otherwise, the continent will fight itself right off the map.[86]

Notes

1. Marc A. Levy, "Is the Environment a National Security Issue?" *National Security* 20, no. 2 (1995): 35–62. See also Daniel Deudney, *The Case Against Linking Environmental Degradation and National Security*, 19:3 (1990) 461–76.

2. Thomas F. Homer-Dixon, "On the Threshold: Environmental Changes as Causes of Acute Conflict," *International Security* 16, no. 2 (1991): 77–116. This theme runs through Homer-Dixon's work. Case studies include Thomas F. Homer-Dixon, "Environmental Scarcities and Violent Conflict: Evidence from Cases," *International Security* 19 (1994): 5–40; T. F. Homer-Dixon, J. H. Boutwell, and G. W. Rathjens, "Environmental Change and Violent Conflict," *Scientific American* (1993): 38–45.

3. Michael J. Edwards, "Security Implications of a Worst-case Scenario of Climate Change in the South-west Pacific," *Australian Geographer* 30, no. 3 (1999): 311–30.

4. Martin Parry, Cynthia Rosenweig, Ana Igleisias, Gunther Fischer, and Mathew Livermore, "Climate Change and World Food Security: A New Assessment," *Global Environmental Change* 9 (1999): S51–S67.

5. Nils P. Gleditsch, "Armed Conflict and the Environment: A Critique of the Literature," *Journal of Research* 35 (1998): 381–400.

6. Norman Myers, "Environment and Security," *Foreign Policy* 74 (1989): 23–41. See also Norman Myers, *Ultimate Security: The Environmental Basis of Political Stability* (New York: Norton, 1993).

7. Jon Barnett, "Destabilizing the Environment-Conflict Thesis," *Review of International Studies* 26 (2000): 271–88.

8. IPCC, R. T. Watson, M. C. Zinyowera, R. H. Moss, eds., *The Regional Impacts of Climate Change: An Assessment of Vulnerability. A Special Report of the IPCC Working Group II* (Cambridge: Cambridge University Press, 1998).

9. V. Percival and T. Homer-Dixon, "Environmental Scarcity and Violent Conflict: The Case of South Africa," *Journal of Research* 35 (1998): 279–98.

10. United Nations Economic and Social Council, Economic Commission for Africa, *Report on Food Security and Sustainable Development (Population, Agriculture, and Environment): Incorporation of Population and Environment Dimensions in Sustainable Agricultural*, 2000.

11. United Nations, *The Least Developed Countries 1999 Report. Conference on Trade and Development*, 1999.

12. Jeffrey Herbst, "Responding to State Failure in Africa," *International Security* 21, no. 3 (1996): 120–44. This view is shared by several authors including Samuel P. Hutington, "Clash of Cultures," *Foreign Affairs* 72, no. 3 (1993): 4–34; Robert Kaplan, "The Coming Anarchy," *Atlantic Monthly* 27, no. 3 (1994): 2.

13. www.onwar.com/aced/data/bravo/bfomli1985.htm.

14. United Nations Integrated Regional Information Network, June 23, 2001; Benin, "Niger Take Dispute to the Hague," distributed by allAfrica.com.

15. Herbert Howe, "Lessons of Liberia: ECOMOG and Regional Peace Keeping," *International Security* 21, no. 3 (1996): 145–76.

16. *Human Rights Reports* Mali, 1993 and 1994; and Niger, 1993 and 1994. http://www.hrw.org/africa/.

17. Abraham M. Williams, "War Crimes Tribunal for Liberian Warlords," http://www.theperspective.org/tribunal.html.

18. BBC, August 10, 1999.

19. Rhoda Howard, "Civil Conflict in Sub-Saharan Africa: Internally Generated Causes," *International Journal* 11, no. 4 (1996): 27–53. Read also Robert Mandel, "Roots of the Modern Interstate Border Dispute," *Journal of Conflict Resolution* 24 (1980): 427–54.

20. Richard Joseph, "Correspondence: Responding to State Failure in Africa," *International Security* 22, no. 2 (1997): 175–81.

21. G. Jan Van Apeldoorn, *Perspectives on Drought and Famine in Nigeria* (George Allen and Unwin 1981).

22. UNOEOA, "Review of 1985 and 1986 Emergency Needs," *Special Report on the Emergency Situation in Africa* (New York: U.N. Office for Emergency Operations in Africa, 1986).

23. Michael J. Watts, "Drought, Environment and Food Security: Some Reflections on Peasants, Pastoralists and Commodization in Dryland West Africa," in M. Glantz, ed., *Drought and Hunger in Africa: Denying Famine a Future* (New York: Cambridge University Press, 1987).

24. Reginald Herbold Green, "Hunger, Poverty and Food Aid in Sub-Saharan Africa: Retrospect and Potential," *Disasters* 10, no. 4 (1986): 288–302.

25. Patrick Webb, "Coping with Drought and Food Insecurity in Ethiopia," *Disasters* 17, no. 1 (1993): 33–47.

26. *Washington Post Foreign Service*, Wednesday, February 7, 2001, p. A12.

27. Jeremy Shoham, "Drought on a Visit to Sudan November 1984–85," *Disasters* 9, no. 1 (1985): 12–14.

28. Kevin M. Cleaver, A *Strategy to Develop Agriculture in Sub-Saharan Africa and a Focus on the World Bank*, World Bank Technical Paper, No. 203, Africa Technical Department Series, 1993.

29. Charlotte Benson and Edward Clay, *The Impact of Drought on Sub-Saharan African Economies: A Preliminary Examination*, World Bank Technical Paper, No. 401, 1998.

30. Timothy M. Shaw and C. E. Adibe, "Africa and Global Developments in the Twenty-first Century," *International Journal* 51, no. 1 (1996): 1–26.

31. Catherine Newbury and David Newbury, "Rwanda in the 1990s: Democratization and Disintegration," in *The Democratic Challenge in Africa*, Discussion Papers 13–14 (Carter Center, Emory University, 1994), 43–52.

32. *Washington Post Foreign Service*, Wednesday, February 7, 2001, p. A12.

33. R. H. Faulkingham, "Subsistence Strategies of Hausa Farmers under Drought Conditions," in M. M. Horowitz, ed., *Colloquium on the Effects of Drought on the Productive Strategies of Sudano-Sahelian Herdsmen and Farmers* (Binghamton, N.Y.: Institute for Development Anthropology, 1976): 32–42.

34. Faulkingham, "Subsistence Strategies."

35. Faulkingham, "Subsistence Strategies."

36. African Newswire Network (ANN), August 25, 2000, www.africanewswire.com.

37. J. Connell, "Climatic Change: A New Security Challenge for the Atoll States of the Southwest Pacific," *Journal of Commonwealth and Comparative Politics* 31 (1993): 173–92.

38. David John Pratt, Fracois Le Gall, and Cornelius Haan, *Investing in Pastoralism*, World Bank Technical Paper, No. 365, 1997.

39. H. Blackburn, H. Glimp, D. Phillipi, D. Child, and A. Pope, "Famine Migration Strategy Paper: Utilization and Preservation of Livestock Resources," *OFDA/ISDA Famine Migration Activity PASA* (Washington, D.C.: USAID, 1993).

40. W. M. Adams and R. C. Carter, "Small-scale Irrigation in Sub-Saharan Africa," *Progress in Physical Geography* 11 (1987): 1–27.

41. K. Kimmage, "Small-scale Irrigation Initiatives in Nigeria: The Problems of Equity and Sustainability," *Applied Geography* 11 (1991): 5–20.

42. A. Tarhule and M.-K. Woo, "Characteristics and Use of Shallow Wells in a Stream Fadama: A Case Study in Northern Nigeria," *Applied Geography* 17 (1997): 29–42.

43. *Post Express*, June 30, 1999. See also ANN, August 17, 2000.

44. *Post Express*, July 7, 1999.

45. *Nigerian Guardian*, July 3, 1999.

46. GRi Newsreel, 1999, www.ghanareview.com/news_subindex.html.

47. IRIN, June 30, 2000, www.reliefweb.int/IRIN.

48. See article by Andrew Furber, "Fulani and Zarma Tribes Pushed into Fighting by Desertification?" *ICE Case Studies*, www.american.edu/TED/ice/nighe.htm, 1997.

49. H. Van Dijk, "Land Tenure, Territoriality, and Ecological Instability: A Sahelian Case Study," in J. Spiertz and M. G. Wiber, eds., *The Role of Law in Natural Resource Management* (The Hague: VUGA, 1996), 17–45. See also D. Van Est and R. Noorduyn, "Resource Use by Floodplain Communities," in H. De Iongh (Chair), *Managing the Dry African Savanna: Options for Conservation and Sustainable Use* (The Netherlands: International Seminar, Center of Environmental Science, Leiden University, 1997).

50. Ruth E. Noorduyn and Woulter T. De Groot, "Environment and Security: Improving the Interaction of Two Science Fields," *Journal of Environment and Development* 8, no. 1 (1999): 24–48.

51. T. F. Homer-Dixon, "Environmental Scarcities and Violent Conflict: Evidence from Cases," *International Security* 19, no. 1 (1994) 5–40.

52. T. F. Homer-Dixon and M. A. Levy, "Correspondence: Environment and Security," *International Security* 20, no. 3 (1995) 189–98.

53. IRIN, June 23, 1999.

54. www.nigeriatoday.com/akande_imposes_curfew_on_ife.htm.

55. BBC, Monday, June 7, 1999.

56. *Lonely Planet*, 2001, www.lonelyplanet.com/destinations/africa/nigeria/.

57. Vanguard, July 12, 2001.

58. Vanguard, July 16, 2001.

59. www.allafrica.com.

60. Gerald Foley, Willem Floor, Gerard Madon, Elhadji Mahamane Lawali, Pierre Montagne, and Kiri Tounao, "The Niger Household Project: Promoting Rural Fuelwood Markets and Village Management of Natural Woodlands," World Bank Technical Paper, No. 362, 1997.

61. United Nations Economic and Social Council, Economic Commission for Africa, "Report on Food Security and Sustainable Development: Development of Xanthan Gum Technology for Enhancing Food Security in West Africa," Fourth Meeting of the Intergovernmental Committee of Experts (Dakar, Senegal, 2001) 26–29.

62. www.bartleby.com/151/a15.html.

63. M. Glantz, ed., *Drought and Hunger in Africa: Denying Famine a Future* (New York: Cambridge University Press, 1994).

64. E. E. Osaghae, "The Ogoni Uprising: Oil, Politics, Minority Agitation, and the Future of the Nigerian State," *African Affairs* 94 (1995) 325–44.

65. *Guardian* (UK), Series in the UK Guardian on the Niger Delta. September 15, 16, 18, 1999, www.seen.org/grdnigser.html.

66. Andrew Rowell, *Shell-shocked: The Environmental and Social Cost of Living with Shell in Nigeria* (Amsterdam: Greenpeace International, July 1994).

67. Geraldine Brooks, "Sick Alliances: Shell's Nigerian Fields Produce Few Benefits for Regions Villagers," *Wall Street Journal*, Friday, May 6, 1994.

68. E. Hutchful, "Oil Companies and Environmental Pollution in Nigeria," in Claude Ake, ed., *Political Economy of Nigeria* (London: Longman Press, 1985).

69. Jim Lee, Kerstin Moesinger, and Amy Maglio, "Ogoni and Nigeria Conflict over Oil," ICE Case Studies, 1997, http://www.american.edu/TED/ice/ogonioil.htm.

70. www.greenpeace.org/~comms/ken/onsla001.html.

71. *Post Express*, November 2, 1998.

72. *Guardian*, July 14, 2001.

73. Vanguard, June 25, 2001.

74. *Post Express*, July 28, 1999, www.postexpresswired.com/.

75. N. P. Sharma, T. Damhang, E. Gilgan-Hunt, D. Grey, V. Okaru, and D. Rothberg, *African Water Resources: Challenges and Opportunities for Sustainable Development*, World Bank Technical Paper, No. 331, 1996.

76. ECA, "Technological Policy Options for Enhancing Food Security in West Africa," *Technical Handbook on Composite Flours*, 1st Meeting of the SRDC/WA, Niamey, Niger, May 25–29, 1998.

77. Irin, July 19, 2001.

78. Reported by Saxone Akhaine and Taofeek Ayinde, www.nigeriatoday.com/violence.htm.

79. Reported by Lekan Sanmi, www.nigeriatoday.com/ad_governors_link_crises_to_enem.htm.

80. IRIN, March 26, 2001.

81. *Nigerian Guardian*, March 20, 2001.

82. K. O'Brien, L. Sygna, L. O. Naess, R. Kingamkono, and B. Hochobeb, "Is Information Enough? User Responses to Seasonal Climate Forecasts in Southern Africa," *Report to the World Bank, AFTE1-ENVGC Adaptation to Climate Change and Variability in Sub-Saharan Africa, Phase II*, 2000, www.cicero.uio.no.

83. FAO, *Africa Special Report*, June 1992.

84. Andrea Gunilla and B. Beckman, "The Wheat Trap: Bread and Underdevelopment in 1985, www.fao.org/WAICENT/FAOINFO/SUSTDEV/.

85. Bureau of Intelligence and Research, 1999, www.state.gov/www/regions/africa/9907_africa_conflict.html.

86. The author gratefully acknowledges the helpful comments and suggestions provided by Professors James Gire (Virginia Military Institute) and Betty Harris (University of Oklahoma) to an early draft of this chapter.

CHAPTER FOUR

Environmental Flashpoints in Africa: Ethiopia and the Blue Nile

John P. Warren

The multiple effects of environmental neglect and degradation in twenty-first-century Ethiopia could precipitate serious domestic upheaval in this ancient nation of sixty million people. Having suffered a series of devastating droughts, decades of border conflicts, the aftereffects of a generation under the oppressive governance of the Derg (the communists who took power after the fall of Emperor Haile Selassie in 1974), and deep cultural and religious differences, Ethiopia continues to struggle with economic development.

Twice the size of France and wonderfully endowed with a unique natural resource patrimony, Ethiopia was long accustomed to having an unending supply of undeveloped land from which to draw its former agricultural bounty. In a twentieth-century scenario repeated in many countries, the natural resource base was overexploited, its management neglected, and environmental concerns were relegated to last place on a long list of national and regional priorities. Now a burgeoning population with growing demands is expanding existing pressures on an environment which no longer possesses the resiliency to respond in the short term. Nowhere are these environmental pressures more intense than in the Blue Nile highlands of the Amhara National Regional State (ANRS) and adjacent areas north and northwest of Addis Ababa.[1]

While such factors as increasing crop failures, poor yields, scarce grazing land, soil erosion, deforestation, and associated human suffering all have near-term implications for domestic governance by the Government of the Federal Democratic Republic of Ethiopia (GFDRE), any mention of the Blue Nile itself is charged with potential international repercussions with respect to Sudan, and especially Egypt. Hence, in Ethiopia as in few other nations, domestic environmental policy and natural resource management actions have the potential for generating immediate international reaction.

Accordingly, it behooves the GFDRE and its local and bilateral partners to move effectively and carefully toward policies and activities that can reverse long-term environmental degradation, improve the sustainable productivity of natural resources, and enhance the well-being of Ethiopians, while maintaining fruitful relationships with neighbors. Otherwise, domestic and regional tranquility would be at risk.

Environmental Situation in the Region

Ethiopia was essentially self-sufficient in food production well into the 1970s, but has failed to meet its own food needs in any year since 1984, and has suffered declining nutritional levels for the past fifteen years.[2] (The GFDRE recognizes the importance of this issue and recently developed a national food security strategy that focusses on increasing food availability, increasing access to food in chronically vulnerable areas, and providing targeted transfers to selected households.) The farmers and stockmen of the ANRS traditionally provided a large share of the nation's agricultural output, but the vast uplands of this region are now characterized by widespread overgrazing, deforestation on a broad scale, and extensive areas of gully and sheet erosion—in some places down to bedrock. The results of these conditions in the Blue Nile watershed include reduced livestock carrying capacity (in an area that places cultural as well as economic value on domestic animals), reduced crop yields and production, lowered family incomes, poverty, downstream flash flooding and siltation, increased vulnerability to drought and associated famine and disease, and increasing pressure for rural dwellers to migrate to urban areas which are ill-prepared to meet their needs.

Causes and Effects of Environmental Problems

Environmental degradation such as that observed in the Ethiopian Highlands can have numerous causes, many of which can be grouped into three categories: (1) population pressures, (2) market failures, and (3) failures of public policies. Ethiopia certainly has experienced substantial population growth over the past few decades, but the destruction of resources cannot be blamed only on that factor, which is common to many countries. A number of market failures can affect resource use and management, including:

- resource owners may not be able to assess individuals or firms utilizing the resource for all the costs which their use imposes;
- resource owners may use inappropriate planning horizons with respect to natural resource allocations, due to need, impatience, or uncertainty;
- property rights may be undefined or absent; and
- market structure for pricing of resources may be weak or nonexistent.

All of these market failures existed to some extent during the seventeen years spent under the centralized economy of Ethiopia's previous communist government. Vestiges of these market failures remain today,[3] and should be considered prime objectives in the context of reformulating public policies. Regarding policy, a central focus for policy analysts and policy makers should be correcting adverse effects of market failures. Of course, there are policy failures also with respect to environmental issues, for example:

- tax breaks or provision of free or subsidized services or infrastructure that underwrite deforestation through support to herdsmen or loggers would be a policy failure, vis-à-vis sustainable resource management, from the point of view of resource owners (often society as a whole);
- the lack of public dialogue involving the participation of resource users and provision for consideration of community rights and expectations in the policy process would be a policy mistake, as it would tend to alienate segments of the population of stakeholders from policy makers (and enforcers);
- failure to provide secure land titles historically has been a common policy failure in many areas, even where definition of real property rights has been an objective.

Such policy failures would exacerbate the negative effects of the market failures enumerated above, increasing the risks for disaffection and domestic unrest.

Domestic Environmental Considerations

During the early decades of the Cold War, Ethiopia under the imperial regime of Haile Selassie was characterized by a strong central government which nevertheless allowed selected provincial authorities varying degrees of autonomy at regional levels. An old system reminiscent of feudalism was slowly moving toward a more open market economy with strong ties to the West throughout the postwar regime of Selassie. A somewhat enlightened view with respect to education, and the emerging market economy were slowly but visibly changing Ethiopia in the 1960s and early 1970s, but the direction of change was altered dramatically with the overthrow of Selassie in 1974.[4]

The ensuing seventeen years were marked by increasing isolation from the West, the embrace of antimarket communist ideologies, nationalization of land holdings and businesses, and attempts at full central planning of the economy. On the natural resources side, forests were bulldozed and forcibly settled by urban dwellers, collective farms were established, and the activities of thousands of Ethiopia's farmers were dictated to them, accompanied by government "recipes" and seed/fertilizer "packages" delivered to farmers through village agents. Without the security of property rights, and with the pressure of production quotas and specified crop selections, the efficiency effects of a functioning market on the agricultural economy are lost, and the incentives farmers have for treating the land well are associated with fear of government reprisal rather than sustainability of soil resources. It is no wonder that the once strong agricultural economy of Ethiopia was devastated by nearly two decades of market—and government—failure.

Today, many scars remain from those difficult years. The present government inherited an economy in shambles. It is not now simply a matter of reestablishing a market economy. Soil erosion has continued to get worse. Overgrazing by herdsmen is widespread. The effects of deforestation are visible throughout the region. Livestock carrying capacity has gone down. Crop yields and productivity have suffered. The agricultural research, education, and extension systems, originally based on the eminently successful U.S. models,[5] were practically destroyed during the 1970s and 1980s and have yet to recover, due in large part to budgetary and policy shortcomings. Decades of economic failure, dependence on central government paternalism and control, and uncertainty regarding land availability have tended to discourage and demoralize farmers and their families, 40 percent of whom reportedly are unable to meet basic nutritional needs.[6] Population growth continues unabated. Ethiopia depends on imports and on food from the international donor community to feed its population.

A serious problem related to environmental degradation in Ethiopia is the fact that the damaged natural resource base is not very resilient in terms of withstanding weather extremes such as drought and flood. This is a common characteristic of damaged ecosystems, and one that tends to worsen without substantial watershed rehabilitation. Recovery from even normal dry seasons or normal rainy seasons can be difficult, as ground cover is depleted and soils are thin and lack fertility. Soil moisture is not retained into the dry season as before, and yields decline. Crop failures become more common.

In Ethiopia today, while collective farms are not the pervasive tool they were under the Derg, farmland remains nationalized. This is another source of frustration and disillusionment for many thousands of rural people. Farmers do not own their land. They work a parcel or parcels assigned to them by the government. Such a system is not conducive to generating investments in the development or maintenance of soil conservation structures, irrigation infrastructure, reforestation or agro-forestry schemes, or other physical improvements aimed at sound and sustainable resource management on agricultural lands.

The land tenure problem apparently is complicated further by instability of parcel assignments due to periodic review and reassignment of rural parcels by the authorities. It would be very difficult to design a system providing more effective disincentives to efficient natural resources management. Farmers are accustomed to uncertainties regarding weather and output markets, but here the farmer is operating under conditions of near-total uncertainty with respect to everything he or she does, including whether or not the parcel(s) being cultivated will be the one(s) cultivated next season or five seasons hence. Farming would be a big enough gamble in Ethiopia without such handicaps. Convincing farmers to modify traditional methods and invest in long-term conservation practices is challenging enough where farmers are assured of secure title to their land, so that the fruits of their labor and investment accrue to them and their families.

Common forests and grazing areas present another challenge to resource management. Of course, the use of common grazing areas has a long history of very acceptable land utilization in many societies. The key to their sustainability, as with any common property resource, is the establishment and enforcement of an often elaborate system of rules. As long as there are more than enough resources to go around, as there were in recent historical times in Ethiopia, the rules are not terribly important. When resource limits are approached, however, the game changes. In a limited common property resource, such as a community woodlot or a fishery, in the absence of regulation, every resource user has an incentive to use as much of the resource as he can, as fast as he can—to "cut down the last tree" or to "catch the last fish," before one of his neighbors does. As the limits of the resource are approached, the incentive to use it becomes progressively more compelling. This natural sequence explains the need in the modern world for natural resources management policies regulating such things as fishing, hunting, logging, farming, irrigating, grazing, and mining of lands, waters, and wildlife seen as public goods. In general, such policies—and their effective application—are missing in Ethiopia.

While its diversity in many areas—including cultural—is part of Ethiopia's wealth and charm, it has also been a source of conflict for many hundreds of years. This nation has many zealous followers of several major religious groups: Christian (primarily followers of the Ethiopian Orthodox Church, closely tied to the Coptics), Moslem, and those holding traditional or animist beliefs. Ethiopia also held a sizable Jewish population, the Falasha, many of whom resettled in Israel in recent decades. The conflict between Christians and Muslims was a major factor in the Ethiopia of the late first millennium, and the heroic defenses of Abyssinia's Christian highlands from the great tide of Islam rolling across northern Africa were the stuff of legend in the European West for hundreds of years.[7]

The Christianity/Islam conflict in Ethiopia today has become more reserved, but nevertheless has important environmental implications. An example is the Great Rift Valley area northeast of Addis Ababa, which abuts the Blue Nile watershed, and where Muslim Afari peoples from the lower, hotter Afari plains (which have become essentially uninhabitable during the dry season) are increasingly coming into contact with the Christian Amhara, the Oromo, and other groups as the Afari bring their herds of cattle, goats, and camels north and west to find dry season water and grazing. Already severely overgrazed, more and more of these areas are being utilized beyond their capacity to recover, and soil erosion and flash flooding are often now the norm when the rains come, rather than rebirth of abundant vegetation. These environmental pressures, spreading along the western escarpment of the Rift Valley, are beginning to spill over into the Blue Nile watershed to the west, which, as we have seen, already suffers from environmental problems of its own. Hence, the prospects for increasing conflicts over land and water use within Ethiopia are further complicated and made more volatile by the cultural and religious differences that may again become more sharply focussed.

International Environmental Considerations

Ethiopia's international relationships present a fascinating historical panorama, from biblical times to the present day. This nation has often been at the center of important international events. Indeed, "World history is being made in and about Abyssinia, whose fate may seriously affect the future of civilization as we know it," as then Field Marshal Viscount Allenby remarked in a foreword to Rosita Forbes' book, *From Red Sea to Blue Nile*, in 1935.[8] That was certainly true in the late 1930s, and in the last third of the first millennium as well. It may have been true in the time of Solomon, and it may be true again, since one of the world's great rivers gets the great majority of its flow from Ethiopia.

From its source at Lake Tana, the Blue Nile traverses some 560 miles of Ethiopia before entering Sudan on its way to joining the White Nile at Khartoum; thence the mighty Nile feeds and waters Egypt. Incredibly, at the outset of the twenty-first century, with the exception of a small diversion for hydroelectric power production at Blue Nile Falls a few miles below Lake Tana, this river flows uninterrupted by dams until it reaches Roseires Dam, fifty miles inside Sudan.[9] It passes on its way thousands of square miles of actual and potential farms and pastures, baking and parched in the dry season heat. Only small-scale diversions on a limited number of upland tributaries have been developed to date, since Ethiopia's attention in recent times has been drawn from natural resource issues to a transition from communist rule, a border conflict with Eritrea, and a series of droughts.[10]

Ethiopia's links with Egypt have been numerous, and frequently positive. For generations there have been strong religious ties between the Coptic Church of Egypt and the Ethiopian Orthodox community. Ethiopia's Blue Nile system (with the tributary Atbara River of Eritrea) produces 85 percent of the Nile's flow.[11] Since the Nile is the only major water source in one of the world's driest regions, any actions by Ethiopia that might intercept significant amounts of that flow could have repercussions downstream that would provoke international confrontations.[12] There is no doubt of Egypt's seriousness with respect to water. Anwar Sadat stated in 1979 that "the only matter that could take Egypt to war again is water," and Boutros-Ghali in 1988 said, "the next war in our region will be over the waters of the Nile, not politics."[13]

Perhaps the most widely reported of Ethiopia's recent international relationships has been the recurring border conflict with Eritrea. It has seemed for a long time that whenever this costly conflict appears to have settled down, one side or the other starts it up again. From observations and discussions in the general area during the mid-2000 outbreak of hostilities, it would appear that chief environmental impacts were related to the torching of opponents' farms and grain stores, periodic movement of human refugees and accompanying livestock, and the expenditure of substantial human and fiscal resources that might have been directed partly to mitigating environmental or related economic problems.

Environmental Security: The Blue Nile Problem

The security risks for Ethiopia which are associated with environmental problems can be considered in two broad areas: domestic and international. On the domestic side, as we have seen, poor stewardship in many forms and for many reasons has progressively devastated the soil and surface water resource base upon which the economy of the Ethiopian highlands depends. The more intensively these resources are used, the less productive they become. The less productive they are, the harder farmers must work them to glean a living for their families, and the more extensively the land area must be utilized. Added inputs such as fertilizers, pesticides, and irrigation often must be applied to coax a yield from depleted soils. These inputs are expensive, often difficult to obtain and distribute, and may cause other problems if used incorrectly.

The current system of land tenure discourages on-farm investment in soil and water conservation practices and infrastructure, and actually encourages overutilization of resources. Concomitant declines in rural incomes and nutritional levels add to the burdens and frustrations of millions of rural people in northern Ethiopia. Many, now accustomed to paternalism and regular delivery of food relief, probably see little hope for improvement. Thousands leave the farm annually to seek a better life in urban areas already crowded with unemployed and underemployed people existing without basic services and without education or resources to improve their lot. It is difficult to envision events that would lead to a rapid reversal of these conditions, and to relief of the hopelessness being experienced by those driven from the land by environmental degradation, and the desperation of those still trying in spite of it to wrest a living from the soil.

Under such environmental and economic conditions, which apparently continue to worsen, ancient tribal, cultural, and religious differences could easily rise to the surface, resulting in local confrontations and incidents of violence. (We need look only "next door," to Sudan, where warfare between ethnic groups has been going on for years, to see that the potential exists for such conflicts.) Even in the absence of cultural sensitivities, collective desperation can boil over into manifestations of anger and antiestablishment or antigovernment sentiment. Individuals intent on causing unrest could easily work such situations to their advantage. Similarly, one more natural disaster, such as a drought or even an unusually dry rainy season, or the failure of donors, transporters, or local authorities to deliver food on time, might be what it takes to trigger domestic unrest.

On the international side, the problem is water. While for thousands of years the rainfall and topsoil of the Blue Nile were the annual gift of Ethiopia to the farmers of Egypt, Ethiopia can ill afford to keep giving the topsoil, which no longer benefits Egypt's farmers, but stacks up behind the dams of Sudan and Upper Egypt. The rainfall is another matter. One answer to Ethiopia's poor productivity in the Blue Nile highlands would be to harness the waters of the river for purposes of irrigating thousands of acres of dryland farms. The problem is, Egypt has continued to develop new agricultural lands and now utilizes about 99 percent of the water that reaches it via the Nile, which does not leave much for Ethiopia to use. Of course, Ethiopia "owns" the source of the Nile, but Egypt has another view, and Egypt is a powerful nation.

In a dry land, water is the issue. Egypt contributes nothing to the flow of the Nile, but understandably considers any upstream development a national security threat. In Sudan, one very good explanation for the long civil war in that country is that access to natural resources, including water, is a major cause of the war.[14] Of course, the case of Sudan is an internal conflict, with strong ethnic divisions. What any large-scale unilateral development of Nile surface water in Ethiopia would probably precipitate would be an international conflict. However, both Sudan and Ethiopia have plans for greater use of the Nile. Coupled with Egypt's plans to continue its developments, it is very difficult to see where the water will come from. In Ethiopia, many small-scale projects remote from the Blue Nile itself but within its watershed will probably continue to be built. In terms of their impact on the flow of the Nile, each is less than the proverbial drop in the bucket, but collectively these projects could become a concern. Certainly, any large-scale irrigation scheme involving damming or diverting the river's flow within Ethiopia would be viewed downstream as a major threat, and a definite risk to peace in the region.

Mitigating the Blue Nile Problem

Were the authorities in Ethiopia to ignore the environmental problems articulated above, the danger of violent internal and external conflict would continue to escalate. However, the environmental risks facing Ethiopia are so extraordinary and overwhelming that to suggest they could all be resolved unilaterally in any short- or mid-term time span would be fantasy. Nevertheless, it is not unrealistic to suggest that the major risks can be mitigated, and that the Blue Nile problem can be "defused" within a reasonable time span of, say, a decade or two. (Actual rehabilitation of the Blue Nile watershed will take longer.) It would require a significant policy shift, a massive development investment, and effective diplomacy by the GFDRE, along with substantial commitments by international donors and nongovernmental organizations (NGO), and international cooperation among neighboring nations and their friends.

Suggested GFDRE Actions

The GFDRE is not unaware of environmental difficulties and their associated impacts on rural productivity and incomes. The Ethiopians are engaging potential international and domestic partners in attempts to ameliorate these problems, but so far, these efforts have overlooked some underlying policy and market factors that will continue to jeopardize sustainable development initiatives. Ethiopia's government—its policy-making elite—should buy into the concept of sustainable development. It should recognize the magnitude of the problem and make a commitment to deal with it, beginning with the underlying market and policy failures of the past. With respect to breaking the vicious cycle that characterizes the situation faced by farmers and their families, the GFDRE should first change its policies regarding land tenure in the highlands. Individuals should be given the incentive to take care of their productive resources, and to put them on a sustainable yield production regime. There should be a functioning market for farmland. Farmers able to purchase and own land are willing to improve and maintain it. The privatization process is a complicated one, especially when it involves so many producers close to the subsistence level, but it is implementable. There are international organizations with experience in this process.

In a country as varied as Ethiopia, there should be exceptions to universal privatization of land. There is a place for carefully regulated common grazing areas, as well. These may be associated with villages, municipalities, family groups, or other associations. In some areas, inhabited primarily by nomadic pastoralists, rather than fixed agriculturalists, there is a strong rationale to maintain a system of common grazing land. However, these types of land management, in the face of increasing populations, should be accompanied by policy determinations designed to achieve sustainability of the resource, by placing carrying capacity limitations, rotating closures, or other parameters on their exploitation.

Importantly, the GFDRE should enlist the cooperation of donors and the NGO community on a huge scale to revitalize their technical assistance outreach mechanisms for farmers and rural cooperatives. There are countless sustainable development technologies available for replication (e.g., terracing, retaining walls, multiuse live barriers, fuelwood efficient cooking stoves, agro-forestry techniques, reduced tillage, crop diversification, water harvesting, post-harvest protection, integrated livestock systems), many of them already in small-scale use in some parts of Ethiopia.

In many areas, an effective operational approach would be to help organize producers around micro-watersheds, some of which would include small-scale irrigation systems. In such a context, natural resource management is a social activity which encourages individual effort for the collective well-being of the community, as opposed to the resource destroying outlook we earlier associated with mismanagement of a common property resource. For example, a village of small holders, all dependent upon the same watershed, would all gain if that watershed were managed to control flash floods and maintain high water quality. That type of management pays important dividends, but requires each individual to manage his or her land effectively. There are plans to pursue this approach on limited sites in the ANRS now,[15] but achieving meaningful mid-term impact would require a massive effort on a larger scale throughout the Blue Nile highlands.

The GFDRE should privatize the agricultural input supply function, and should encourage and facilitate the operation of free markets for both inputs and the products of agriculture, agro-forestry, animal husbandry, and rural industry. This would free up public funds and encourage growth of the private rural economy, allowing markets to perform effectively.

Fundamental change is implied in these suggestions, and that kind of change over a broad geographic area will take planning, commitment, investment, effort, and time. There is no quick fix to a problem that has been so long in the making. Existing programs, especially including emergency relief efforts, would have to continue in the interim, until replaced or overtaken by events. In spite of every effort, some farmers will fail. That is part of the free market system. However, results in other areas suggest that investments in soil conservation almost universally have a very positive benefit/cost ratio, that land ownership has been a critical factor in transforming traditional practices and establishing sustainable production, and that farmer incomes frequently show significant near-term increases associated with participation in group training in agro-forestry and sustainable agriculture practices.[16]

As part of a broad sustainable development initiative, new or revised GFDRE policies may be needed to provide guidance on numerous environmental issues relating to pollution, contamination, overutilization of vegetative cover, deforestation, erosion control, and related factors. Public sector agencies with environmental and/or natural resource management responsibilities may require upgrading of their capabilities in this regard.

On the subject of Nile River water diversions, it is unlikely that the near term will bring any major surface water developments in Ethiopia. It is more likely that small-scale systems in the upper parts of the watershed will continue to be designed and built. However, it is probable that larger schemes are in Ethiopia's future, and the planning for such activities will require some international brokering if conflict is to be avoided. Any major diversion of Blue Nile waters by Ethiopia would require adjustments in Nile water use by Sudan, Egypt, or both. As the final user, Egypt may not be willing or able to make major adjustments, especially in the near term. Egypt is not only developing new Nile water schemes, but is also developing groundwater resources in the desert. These ancient aquifers may no longer have natural recharge, so there may be a future reckoning on water supply that will be very challenging, indeed. Perhaps there are underutilized options for water conservation in Egypt. Perhaps desalinization will be an economical solution by that time. In any event, indications are that the Blue Nile may be entering a very contentious phase.

There is reason to be somewhat optimistic in this regard, in spite of the apparent dangers, and the pronouncements of Egyptian statesmen regarding the possibility of war over the waters of the Nile. It has been reported that the last war actually fought over water was between Mesopotamian city-states 4,500 years ago, although there were seven minor skirmishes over water in the twentieth century.[17] Nevertheless, prudence would require a very careful approach to surface water developments affecting the Nile's flow, if such development is to be achieved in a cooperative process. Negotiations relating to such matters would be time-consuming and sensitive, and would undoubtedly require diplomatic participation and probably arbitration by third parties and/or international bodies.

Ethiopia's Development Partners

The government of Ethiopia has access to numerous intergovernmental, financial, multilateral, voluntary, commercial, religious, humanitarian, and educational partnerships. Examples include government to government bilateral economic assistance programs from European and Asian nations, Canada, and the United States; international voluntary organizations like Save the Children, CARE, Doctors Without Borders; multilaterals such as the United Nations and the World Bank; local NGOs and community groups; universities; and religious organizations. Many of these entities have been at work in the country for years, and have developed strong ties among the people and institutions of Ethiopia. Health care, emergency food relief, primary and secondary education, physical infrastructure, and agriculture are a few examples of existing development partnerships.

Traditionally, the environment and long-term management of natural resources, while fundamental to sound economic development, understandably have had lower priorities than providing basic health and nutritional requirements for human survival during drought, famine, and displacement. An unfortunate fact is that environmental degradation on the scale experienced in Ethiopia contributes to drought, famine, and displacement.

Perhaps it is time for appropriate partners, with the GDFRE (civil and military branches), to organize an environmental council or task force to examine security aspects of the environmental situation, and to develop a strategy for the involvement of interested partners in a plan to mitigate the potential Blue Nile problem in the environmental context. The problems are too daunting for unilateral solutions, too pervasive to be ignored, and too complex for superficial treatments. The environmental issues in the region require immediate, serious attention, and a long-term concerted effort to address them would have important and enduring impacts on many of Ethiopia's other development problems, as well.

Conclusions

Ethiopia at the beginning of the twenty-first century is an ancient land with bountiful resources, strategic location, a richly fascinating history, and a wealth of potential for its future. That potential, however, is jeopardized by the worsening effects of natural resource degradation on a massive scale throughout large areas of the country. Environmental damage is robbing the Ethiopian people of many things. One of the most important of those things is the ability of a proud rural people to produce the food and fiber it has historically provided—and the independent livelihood Ethiopian families enjoyed for generations. Resource deterioration is also robbing millions of Ethiopians of their health, their productivity, and their hopes for a prosperous and stable future. Indeed, it threatens domestic tranquility and regional peace between their homeland and its neighbors in northeastern Africa.

There are many factors that have led to this state of environmental insecurity, and many paths to take in attempting to improve conditions. If the temptation arises to "sweep away" all these problems by undertaking, for example, a grandiose infrastructure project such as placing a dam on the Blue Nile, the Ethiopians would be well advised to resist it. At the least it should be approached with great care, and should be only one aspect of a fully integrated effort. On the one hand, such a project could precipitate international conflict with Egypt, and on the other, it would still ignore the root causes of the environmental devastation ocurring in Ethiopia. The environmental damage in Ethiopia has little to do with the Blue Nile itself, and everything to do with how people treat the resources of the Blue Nile highlands (and other areas). Building a dam will not change those underlying problems. A dam built on the Blue Nile today will be full of silt "tomorrow" unless the root problems are addressed.

Ethiopians have not devastated the Blue Nile highlands because they set out determined not to do so. It has happened due to population growth and poor agricultural practices that still worked as long as there was excess land. The farming and grazing and forestry practices in use for generations should have been improved and adapted to changing conditions, through extension training, farmer organizations, and market pressures. Those practices were not modified, due to the failure of and/or absence of functioning markets (for land, for soil, for water, inputs, outputs) and of effective government policies. Until these fundamental issues are dealt with, and a sensible policy regime is in place, matters will continue to get worse. Mitigating the immediate problem of watershed degradation is possible, but it requires commitment, investment, concerted effort, and time. The willing participation of international partners is undoubtedly available, but it is up to the Ethiopians to take the first steps and mobilize the response that this environmental flashpoint warrants and demands.

Notes

1. Elements throughout this presentation were drawn from the author's observations during periods of residency in Egypt, Ethiopia, and other countries, and a two-month temporary assignment to Ethiopia in mid-2000.
2. John Warren, "Ethiopia and the Transition from Point Four to USAID," presentation at dedication of Point Four Room (Stillwater, Okla.: Wes Watkins International Center, Oklahoma State University, October 6, 2000), copy on file with the author.
3. Daniel Taferra, "Ethiopia: Post-Cold War Market Reform and Globalization," draft paper (Ferris State University, Mich: Department of Economics, 2001), copy on file with the author.
4. Taferra, "Ethiopia."
5. See, generally, Oklahoma State University, "Oklahoma State University in Ethiopia: Terminal Report 1952–1968," Stillwater, January 1969.
6. See, generally, Warren, "Ethiopia and the Transition from Point Four to USAID."
7. See Mohamed Amin, Duncan Willetts, and Alastair Matheson, *Journey through Ethiopia* (Nairobi: Camerapix Publishers International, 1997).
8. See foreword by Field Marshal Viscount Allenby in Rosita Forbes, *From Red Sea to Blue Nile: A Thousand Miles of Ethiopia* (New York: L. Furman 1935).
9. See Virginia Morell, "Blue Nile," *National Geographic* 198, no. 6 (December, 2000).
10. See Stephen D. Kiser, "Water: The Hydraulic Parameter of Conflict in the Middle East and North Africa," Institute for National Security Studies Occasional Paper 35, Environmental Security Series, Colorado, September 2000.
11. John H. Spencer, "Ethiopia at Bay: A Personal Account of the Haile Selassie Years," Michigan, 1984.
12. See, generally, Kiser, "Water."

13. See Peter Gleick, "Water Conflict Chronology" (September 2000 version), Internet, www.worldwater.org, accessed May 2001.

14. See Kiser, "Water."

15. See Warren, "Ethiopia and the Transition from Point Four to USAID."

16. See, for example, John Warren, "Improvements in Natural Resources and Environmental Management in Honduras, 1976–1995," prepared for the U.S. Congressional Budget Office Study on the inpact of development assistance, USAID/Honduras, Tegucigalpa, 1996.

17. See A. T. Wolf, "Water Wars and Water Reality: Conflict and Cooperation along International Waterways," NATO Advanced Research Workshop on Environmental Change, Adaptation, and Human Security, Budapest, Hungary, 1997.

CHAPTER FIVE

Security Implications of Latin America's Environmental Problems

Frank McNeil

Oddly, there is a greater recognition now in Latin America and the Caribbean than in the United States of the role of environmental change in spurring conflict and lesser tensions, within and among states.

Though the idea of environmental security is by no means digested within the region's political class, there is growing interest, not just among civil society and policy experts concerned with conflict resolution, but in some instances, among senior civilian and miltary figures. Perhaps they (and similarly minded counterparts in Southeast Asia) are trying to tell us something. It would not be the first time that the United States was slow to recognize new forces which could affect the peace. These people are living the problem, experiencing firsthand the consequences of environmental change, which is why they have taken to environmental security.

Regional observers, particularly from the Caribbean, are quite mindful of at least one "planetary" concern: global warming. But what "concentrates the mind," in Samuel Johnson's phrase, are the local consequences of environmental damage upon societies and, in some instances, relations with neighboring states.

For people in authority from Mexico down through the Andes, their first concern is with the massive, sudden environmental insults known as natural disasters, earthquakes, hurricanes, and volcanic eruptions. Though it is not widely known, parts of the region, even the Caribbean, are also prone to the odd tsunami, e.g., the long ago obliteration of Jamaica's Port Royal.

No year goes by without one, often two or three, major disasters. For small nations, a major disaster overwhelms the capacity of the state to provide for the general welfare and harbors the potential to create serious instability. Disasters also have international ramifications, spurring migration, pushing people on the margins into crime, and placing demands on the international community for help.

No wonder then, that a recent three-day international meeting on environmental security in San José, Costa Rica, which mixed senior environmental and military officials from Mexico, the Caribbean, Central America, Panama, and Guyana with civil society and policy specialists, put a high priority on disaster relief, reconstruction,and mitigation.

The conference, in which U.S. officials and nongovernmental Americans also participated, dealt with other matters, too. They ranged from the mundane, e.g., environmentally sound port management and environmental remediation, to issues on a grand scale, the exhaustion of fisheries and the risks to societies (and to regional comity) from accelerating ecosystem damage. A similar meeting is slated for late this year in Asunción, for South America.

In an earlier, exclusively nongovernmental workship, also held in San José (a convenient site for regional meetings) the principal concentration lay in assessing the links between environmental change, cumulative as well as sudden, and societal and international tensions. All the participants, save the author, hailed from the region, defined as Mexico, the six nations of Central America, plus Panama.

This chapter is about Latin American cases, past, present, and prospective. It relies upon the work of the North-South Center at the University of Miami, in an evironmental security project which has entered its third year, involving many discussions with knowledgeable people from Latin America, and the Caribbean, including the conferences briefly described in the preceding paragraphs.

In looking at these cases, no attempt was made to assign factor values (e.g., percentages) to environmental change as a causative agent in specific conflicts and situations of lesser tension. Recognizing that wars have many fathers, we used what I call the "Kalimantan Test," i.e., does or did environmental change play a significant role in the case in question?

The last section recounts two experiments in case-based forecasting jointly conducted by the Center and the university's Rosenstiel School of Marine and Atmospheric Sciences, which dealt with quite different regions, South America's Pantanal and the "Wider Caribbean," including Central America.

Haiti: The Price of Deforestation

Haiti is the most obvious case in the Western Hemisphere in which cumulative environmental devastation has been a major cause of violence—some say the major cause, though Haiti's peculiar institution of caste distinctions, based on shadings of skin color, cannot be ignored in the accounting.

Haiti's forests were denuded in the last century and have stayed that way through all the brutal regimes and bursts of anarchy. The effects of recent deforestation on watersheds are well known, particularly in terms of its multiplier effect on the devastating floods which attend most major hurricanes in the Caribbean and Central America.

Worse yet, having gone without forests for so long, Haiti lost most of its topsoil, washed down to bare rock by a century of floods, unimpeded by forest cover. For all but a few plantations, hardscrabble farming is all there is to Haitian agriculture. Deforestation has kept Haiti permanently poor, spawning internal violence and tensions with the Dominican Republic, which shares the island of Hispanola with Haiti.

In recent years, abject poverty and rampant instability have spurred numbers of migrants, mostly to the United States Haiti's status as a failed state—and the waves of migrants risking the passage to the United States (no one knows how many have drowned in the trying)—drove the Clinton administration to intervene and the United Nations (UN) temporary stewardship of the island.

For a variety of reasons, internal and external, these interventions have failed, and Haiti is no better economically and only somewhat better politically than during the era of the Duvaliers. Without a major effort to reforest and to restore arable land, no easy task with so little topsoil, Haiti is likely to remain a failed state for the foreseeable future.[1]

Guatemala: The Peten, Then and Now

In 1964, four Americans, a CARE representative, a graduate student, a journalist, and I, took a DC-3 to Sayaxche, a tiny outpost on the Rio de la Pasion in Guatemala's vast Peten territory. From there, we spent a week on the river and its tributaries, traveling by cayuco to archaeological sites, among them Seibal and Altar de los Sacrificios. We lived on rice, beans, and fish from the river. Except for an abandoned CIA camp for the Bay of Pigs where a guard fired a few shots our way, we saw nothing but unspoiled rivers, trackless forest, and ghostly reminders of the Maya.

Gordon Willey, the most distinguished Mayan archaeologist of his time, let us camp at Altar de los Sacrificios with his team in their first season of exacavations (the famed "altar" still rested where it had lain for centuries). Willey had begun to shed new light on a great mystery of the ancient world: what caused the Mayan civilization to disappear, centuries before the arrival of the Spanish.

For one thing, he pointed out, strife had come to the supposedly peaceful Mayans; the stela at Seibal displayed warriors, presumably from Mexican centers to the north, in warlike postures and ceremonial battle dress. Did the gods fail the Mayas? Other discoveries hinted at crop failures, which perhaps undermined their calendric religion, meant to assure the succession of the seasons and abundant foodstuffs. Today, new evidence suggests a drought, a vast environmental insult that endured for nearly two centuries, during the time the Mayan ceremonial centers in Guatemala, Belize, and southern Mexico were abandoned.

At this writing, the Peten still harbors large tracts of primary forest, but land pressures and a bloody civil war have made significant inroads, bringing roads far into the Peten and stressing the ecosystem in various ways. Environmental change did not foster conflict there. Instead, the civil war spread to the Peten, where instability helped drive environmental damage.

In this sense, the Peten is an illustration that environmental security can be a two-way street. Not only can environmental change foster conflict and instability, as it may have done among the ancient Maya, but in certain circumstances, if a war spreads far enough, it can bring about significant ecosystem change.

Shaking the System: Earthquakes and Regime Change

Many observers, including me, date the beginning of the end of the Somoza regime to the great Managua earthquake of 1974, when mismanagement and corruption—particularly on the part of Somoza's National Guard—hampered the international relief effort.

Blood supplies were left to spoil in the sun or stolen and sold. Food supplies were stolen and sold on the market. The corruption, amply reported in the American press, angered the suffering populace and an international community, which had paid little attention to Somoza's depredations prior to the disaster.

From this time forward, the Sandinista guerrillas began to gain popular support against what was previously thought to be an impregnable dictatorship. To be sure, underlying factors were more important than the earthquake itself, above all, the cumulative toll of years of corruption and oppression on the personal legitimacy of the rule of the Somozas, father and son.

Ideology and the Cold War also played important, though lesser, parts in what was, after all, a popular revolution against the Somoza dictatorship supported by many average Nicaraguans who would later vote in large numbers to oust the Sandinistas and install a democratically elected president, Violeta Chamorro.

In like manner, foreign support for the revolution came not only from Cuba and, eventually, the Soviet Union, but also from noncommunist governments in Venezuela, Panama, and Costa Rica, nations which would become disenchanted with the Sandinistas in power. And in the late 1970s, the United States bumbled its chances to persuade Somoza to leave before the course of revolution was set in blood, a fact amply documented by former Carter administration officials, notably Robert Pastor and Anthony Lake.

None of this gainsays the fact that Nicaragua is a clear case of where a sudden, massive environmental insult, in this case an earthquake, helped set in motion a violent change of regime. Something similiar could have happened in Guatemala, when in 1975 a more powerful earthquake struck Guatemala City and its environs, killing thousands of people. But it did not.

The regime, headed by President Kjell Lagerud, a former general, handled the relief effort with considerable dedication. Of course, Lagerud's regime was in many ways an exception to the prevailing brutishness of most of Guatemala's military led regimes, before and afterwards.

In El Salvador, a civil war, lasting from the late 1970s until the early 1990s, exhausted the country, when the Central American peace process finally brought it to an end. In these circumstances, the recent earthquake was all the more devastating, not only physically, but emotionally. Still, the people pulled together, in good part because they want no more war. In particular, the military in El Salvador—responsive to elected leaders—also did a good job with respect to disaster relief in collaoboration with United States and international relief agencies. In the next elections, it is possible, though not certain, that the people of El Salvador will choose to oust the conservative ruling party from government. If so, it will be because the people want new faces and new ideas to confront the challenges facing the country, not just because of the earthquake.

What do three devasting earthquakes tell us about the links between sudden environmental change and instability? In Nicaragua, the earthquake undermined regime legitimacy. In Guatemala and El Salvador, the earthquakes did not, even though Guatemala was in the throes of a lengthy civil war (the bloodiest of all the Central American wars) and El Salvador was trying to recover from its own deadly conflict.

There were other factors, but it is clear that the response to the disaster of the governments of the day (and their respective militaries) was important. Guatemalans consciously kept the example of Nicaragua in mind. It was widely known that General Lagerud told his government and military not to behave like the Somoza regime, under threat of severe punishment.

The "Soccer War"

The literature, including a recently released U.S. Agency for International Development (USAID) analysis from 1986, suggests that the 1966 "soccer war" between Honduras and El Salvador was derived in part from the cumulative effects of environmental degradation. This requires explanation. El Salvador was (and is) a small, overpopulated country where there is nowhere near enough land to go around. The destination of choice is the United States for economic and, until recently, political refugees (here defined as people who fled the war, even if they do not fit the United Nations technical definition of political refugees).

In the 1960s, however, Honduras was the destination for land-poor Salvadoran peasants living near the border. They had no land or their land had been exhausted under the slash-and-burn (roza) agriculture practiced from early pre-Columbian times in Central America. The scarcity of land and overpopulation were, so the analysis runs, environmental causes of the "soccer war," so-called because a riot during a soccer match between the two countries sparked the hostilities.

This analysis, however, is not entirely satisfactory from the perspective of the "Kalimantan test," which stipulates that environmental changes must in some way play a significant role in spurring tensions. To be sure, the Brundtland Commission report cites overpopulation as an environmental threat, and so it is. But while population pressures contribute to environmental damage and to conflict, it is hard to make a case that ecosysystem change is a cause of overpopulation.

Moreover, tensions between El Salvador and Honduras go back a long way, to Central America's independence from Spain in 1821, particularly to the strife between anticlerical "liberals" and pro-Church conservatives which littered much of Central America with bodies for a century.

There is no wish to quarrel with those who hold that the "soccer war" was in part derived from environmental causes, in the "planetary" sense which the literature gives to overpopulation. But population pressures and the other tensions which led to the "soccer war" are classical sources of conflict, the existence of which long antedates most of the environmental changes at issue here.

The preference here is to treat environmental security more restrictively, by confining it to readily demonstrable linkages between environmental damage, mostly but not always, man-made, and conflict or tensions short of war. By this standard, the "soccer war" fails the Kalimantan test. The most one can say is that when the peasants ran out of Salvadoran land, the next stop was Honduras, a phenomenon which began years before the "soccer war."

Today, though no one says such things officially, in Honduras and El Salvador there are a few people who privately predict another clash in about a decade, for much the same reasons as the "soccer war." This is disturbing but unpersuasive as matters stand today.

The United States, not Honduras, is now the destination for migrants from El Salvador. And once rancors from the war had cooled under an international regime supervised by the Organization of American States, the World Court removed another cause of the conflict by settling an age-old border dispute.

Presumably a return of serious tensions between El Salvador and Honduras would require some new factor. Could environmental damage, which is already contributing to nonviolent tensions in several parts of Central America, be that factor, here or elsewhere in the region? The answer, discussed later in this chapter, is a qualified "yes."

Making Distinctions

Of the preceding four historical examples, only one, the Managua earthquake, clearly meets the standard—a case in which environmental change significantly contributed to conflict.

If one asserts, wrongly, that environmental change equals conflict, environmental security is drained of useful meaning, not only in a theoretical sense, but also for the purposes of policy. Not every conflict has an environmental cause, even though conflicts and tensions in some measure derived from environmental change are on the upsurge. It is important to discriminate, and Central America, where conflicts have sometimes, but not always, shown environmental linkages, is a good place to evoke this distinction.

> In the literature, there is criticism, prominently identified with the political scientist, Daniel Dudney, of the idea that one can "prove" a causal relationship between environmental change and conflict. The criticism, directed against the work of Homer-Dixon, is complex, and most of it is not germane to a work which deliberately sets aside theoretical argument. But the charge that proponents of environmental security ignore examples that disprove a relationship between ecosystem damage and conflict ("null cases," as academics style them) may be relevant. Citing a few cases which fail the "Kalimantan test," because links between environmental change and the conflict in question are tenuous, may help address this concern.[2]

Colombia

In Colombia, a reciprocal relationship between the drug trade, civil war, and ecosystem damage has taken on the characteristics of a vicious cycle in which environmental damage puts even greater pressure on a war-torn society.

The strife began in 1948, long before environmental change became a serious issue for Colombia, with the murder of a popular presidential candidate, which was followed by the Bogotazo uprising, a Paris commune-style revolt in Colombia's capital city. Inspired by inequity and ideology, the Bogotazo failed. Nonetheless, it sprawned a half century of endemic insurgency and terrorism.

Since then, civil war has waxed and waned. In the late 1990s, it reached a new peak as profits from the drug trade swelled the coffers of the insurgencies, the left-wing guerrillas and the right-wing militias alike, men without pity who are tearing apart Colombia.

For years, toxic chemical runoffs from cocaine labs have polluted ground water and nearby habitat. The labs are moved frequently, so that by now there are probably thousands of abandoned sites, which by U.S. estimates have released a cumulative total of around five-and-a-half million gallons of precursor chemicals and by-products into the nearby environment.

Now that industrial farming of the coca plant has shifted northward from Bolivia and Peru to Colombia, numerous coca plantations, usually hidden in the jungle, damage Colombia's forest. Aerial surveys indicate that coca growers have had 94,000 acres under cultivation in the prime coca growing province of Putumayo. That takes a lot of pesticide, which the industry can afford, and the estimates are no less plausible than they are astounding.

The controversial aerial eradication campaign, in which crop dusters spray a glycophosphate herbicide on the coca fields, is causing some damage to the nearby environment, because of unintended oversprays and sudden wind shifts. The official U.S. position seems to be that the pilots always hit their targets, even under fire, which should qualify them en masse for the Blue Angels. One could make a far more powerful case for spraying by contrasting the massive and persistent environmental damage from the coca plantations with the limited quantities and effects of glycophosphate.

That said, this chapter is not about the drug war per se, but about the multiplier effect of civil war and the drug trade upon the environment in Colombia. No matter the reader's view of the proper strategy for drugs, he or she should understand that the links between environmental damage and conflict have become a two-way street, in which each feeds the other.

In Colombia, where the roots of war long antedate the existence of an environmental dimension to the conflict, there has grown up over time a reciprocal relationship between war and environmental damage. This relationship has been exacerbated by the guerrillas and the "paramilitary" militias, who force peasants, who must grow their food somewhere, to flee from place to place.

Rivers of Contention

The Pantanal

The Pantanal is the world's largest wetlands, thirteen times larger than Marjorie Kinnan Rawlings' "river of grass," the Everglades. Technically known as the Upper Paraguay River Basin (UPB)/Pantanal region, the wetlands, which originate in the glacier-fed streams of the Bolivian Andes, feed the Paraguay River.

Four countries, Brazil, which has most of the wetlands, Bolivia, Paraguay, and Uruguay, share this watershed. No one will repeat Paraguay's nineteenth-century mistake and attack Brazil, but there is conflict of another sort among local and national stakeholders over the proper uses of the Pantanal.

Contention among stakeholders over the use of major environmental assets is a prominent feature of the Latin American landscape. The Pantanal is a huge example of this insufficiently studied phenomenon, which is not generally seen as a security concern unless and until these tensions lead to a resort to arms.

Without putting too fine a point on the matter, perhaps it is useful to visualize a spectrum of conflictual relations ranging from organized violence to lesser but important tensions, where the area is large enough and the contention, though short of war or rebellion, is significant enough to affect the direction of society.

Often, the key issue for Latin and Caribbean societies is how best to reconcile stakeholder interests in envioronmental assets within a framework which allows for their wise use. Otherwise, major stakeholder conflicts become a drag on progress, a source of corruption (e.g., illegal logging and fishing or illegal toxic waste disposal, winked at by the authorities) leading to localized insecurity.

In the Pantanal, biodiversity loss already has an impact. The stakeholders, cattle, mining, and soybean interests, communities, provincial authorities, and civil society institutions, among them, indigenous groups and environmentalists, vie with each other over how much destruction is enough. The "who gets what" involves not only profitable activities, but also health, water quantity and quality, and damaging floods. Should the Pantanal shrink too far, scientists tell us a disastrous climate change may happen.

Central America

From Mexico's southern frontier to Panama, Central America hosts thirty-seven "international" rivers, large and small. Some, like the Usumacinta and the San Juan, are boundary rivers for much of their course, marking the frontier between Mexico and Guatemala and Nicaragua and Costa Rica, respectively.

Most of them, like the Rio Lempa, begin in one country and end in another, in this case El Salvador and Honduras. All these rivers harbor actual or potential upstream-downstream problems, that is to say, arguments over water usage, pollution, and silting.

From ancient times, the successful management of rivers has been a collaborative affair. Unfortunately, territorial disputes outlast empires and ideologies. Latent or actual disputes over river frontiers complicate watershed management and appear to exert a multiplier effect on environmental damage.

This is true of the dispute over boating rights on the San Juan River, which has paralyzed incipient arrrangements for Costa Rican-Nicaraguan cooperation in managing this watershed. In contrast, there is anecdotal evidence that communities along the Usumacinta are working together, free from central government meddling, largely because there is no residual dispute over the boundary.

Though Nicaragua and Costa Rica are small countries, the San Juan watershed is a major part of what, by some acounts, is the largest single stretch of relatively undisturbed tropical rain forest north of the Amazon basin. Like the tensions between Honduras and El Salvador, contention between Nicaragua and Costa Rica has its origins deep in history.

In the late 1830s, the United States sent its first envoy, John Lloyd Stephens, to what was momentarily a single nation, the Republic of Central America, which disintegrated days after Stephens presented his credentials. The first great archaeologist of the Maya, Stephens later wrote of Nicaragua in his *Incidents of Travel in Central America and Yucatan* that everyone "warns you against them." Stephens had just come from Costa Rica.

The Rio Lempa is an exemplary case of upstream-downstream contention. Without getting into detail, the major risks to the river stem from water takings and pollution and from agricultural and industrial uses; neither upstream El Salvador nor downstream Honduras has done well at river management.

In a worst-case scenario involving the return of serious tension between El Salvador and Honduras, the Rio Lempa is the most likely candidate for that new factor that could play a role in rekindling the risk of conflict. In the words of a highly respected former foreign minister from another country, "The Rio Lempa is a good example for environmental security. It shouldn't be allowed to run dry, but the governments by themselves will never find a solution. Civil society will have to take the lead."

Hurricane Mitch

The hand of man is exacerbating natural disasters. Deforestation, for example, significantly reduces the capacity of the soil to hold water by about a third, according to some scientific estimates, making conditions more favorable for terrible landslides and floods from hurricanes.

This is what happened when flash floods and huge landslides triggered by Hurricane Mitch devastated almost all of Honduras and parts of Nicaragua, leaving about 10,000 dead and countless homes and livelihoods destroyed.

The costs of the destruction are far reaching. Estimates suggest Honduras' economy has been set back anywhere from a decade to a generation. The two countries will receive in total about $2 billion in aid for relief and reconstruction from the international community, about half of that from the United States.

Numerous refugees went elsewhere. From Honduras they headed to the United States, often risking their lives on the "coyote conveyer belt" which runs through Mexico. Nicaraguans sought refuge in Costa Rica, doubling the earlier wave of immigration, sparked by the Central American war. Today, Costa Rica harbors about 500,000 Nicaraguans, about an eighth of Costa Rica's four million people, a fact which has contributed to the tensions between the two countries.

Without deforestation, Hurricane Mitch would have still been a major disaster, but deforestation made it into one of biblical proportions. Spurred by Mitch and earlier disasters, international agencies, including AID's Office of Foreign Disaster Relief, have expanded their focus to include mitigation, defined here as action to reduce vulnerabilities.

Mitigation involves many things: enforcement of building codes for earthquakes, resiting of structures and even communities away from flood plains, and conservation and reforestation of key watersheds. Action on this front will save lives, lessen devastation, and reduce the costs to the economy.

The marriage of deforestation with a major hurricane is an obvious security problem for the "Wider Caribbean." There is, however, no military solution for mitigation, which requires both effective policy and substantial capital. This is a civilian responsibility, unlike disaster relief per se, where regional militaries, now directed by elected leadership, bear a heavy burden.

In Latin America and the Caribbean, the next disaster is always just around the corner. Reducing the vulnerabilities to future disasters is a national task. It ought to be a major priority for international donors, for whom investments in mitigation will have a payoff, over time, in reducing the costs of disaster relief.

Rivers Run through Them

The examples discussed above are striking. Together with the Panama Canal watershed, discussed in the next chapter, they make a persuasive case that river and watershed management is a significant security issue for Latin America and the Caribbean, with international ramifications.

Throughout the region, there are many other "international" rivers not discussed in this chapter. Frequently they have upstream-downstream problems, often involving contention among stakeholders. Most will never make the headlines. Collectively, however, the management of these rivers and watersheds will have a major say in the region's future.

The Pantanal and the "Wider Caribbean" receive further treatment in the last section of this chapter, which deals with an experimental forecasting exercise.

The State of The Sea

One of the world's great fisheries, the Caribbean has been a major source of protein from the east coast of the United States, down through the islands and Central America, to the Caribbean coasts of Venezuela, Colombia, and Panama.

The fishery's annual ex-vessel value is estimated at several billion dollars. The fishery is also tied to other services and activities, among them tourism, which in the aggregate exceed $100 billion a year.

Chronic overfishing has depleted nearly every commercial stock. Near shore ecosystems, among them mangroves, sea grass beds, and coral reefs, suffer from coastal pollution. These ecosystems have reached the point at which human and natural causes are acting together to produce severe habitat and resource degradation. The islands, "small island developing states," in the U.N. definition, are especially vulnerable but so are other coastal regions.

Today coastal zones are at risk from explosive population growth and the sorts of development that pollute the environment. Sewage and industrial effluents have harmed important reefs all around the Caribbean.

Serial overfishing, particularly by extraregional, high tech industrial fleets (where each ship's fishing power has quadrupled in a decade), has driven down fish stocks, even among pelagic fish. One after another, big fish in the reefs, bays, and fringing seas have disappeared, leaving smaller fish with lower reproductive value, threatening the sustainability of the species.

This is what science tells us. What are the results? Fish stocks have declined to critical or near critical levels in many locales. Local, artisanal fishermen are nearly out of business. Less local fish means higher prices, less revenue, and less protein, which, if replaced, is often imported at higher cost.

Maritime disputes of one sort or another complicate matters, e.g., Trinidad/Venezuela, Nicaragua/Colombia, and Honduras/Nicaragua and, on the Pacific side, the Gulf of Fonseca among Nicaragua, El Salvador, and Honduras. International fisheries regimes are almost nonexistent.

Environmental Health

In Latin America and the Caribbean, environmental change is spurring a resurgence of malaria, the spread of dengue, and the transmission of intestinal illnesses, such as cholera and various dysenteries.

In the case of waterborne diseases, coastal pollution and polluted drinking water systems are the culprits. Encroachment, if that be the word, of human settlements into the favorite habitats of mosquitos appears to be a major factor in malaria and dengue.

Mining, industry, and agriculture are necessary occupations. Inadequately regulated, however, they become, along with urbanization, important sources of pollution. Many places in the region suffer health hazards from high levels of sewage, pesticides, solid wastes, and even heavy metals.

Most of the great urban centers of Latin America are incubators of pollution. For years Mexico City's air quality has made that of Los Angeles look positively benign. Waste disposal is a huge problem even for relatively small cities, like San José, the capital of Costa Rica.

There is another way, which Curitaba in Brazil has pioneered, to the point that Curitiba and Singapore are paired as model cities in the developing world. That said, it is far harder for Rio de Janeiro to clean up from the way it is today than for Curitiba, which has led the way in measures to prevent pollution.

Will diseases and pollution stimulate conflict? Not by themselves, probably, though masses of people, often poor, crowded together in massive cities do not provide a recipe for political stability, much less for personal security. What is certain is that disease and pollution levy high costs on fragile economies, place more stress on health care systems, and markedly lessen productivity.

Case-Based Forecasting

During the last half of 1999, the North-South Center and the Rosenstiel School of Marine and Environmental Sciences conducted an experiment in science driven, case-based forecasting, with the purpose of making environmental security more intelligible to policy makers and analysts and to concerned citizens.

The cases dealt with two very distinct regions in the Western Hemisphere, the "Wider Caribbean" and the Upper Paraguay River Basin/Pantanal complex. The North-South Center and the Rosenstiel School designed the studies, which were carried out by teams from the two institutions, joined by scientists and experts from other academic and research instituions.

Three previous sections in this chapter, "rivers of contention" (among them the Pantanal), the great Caribbean fishery, and environmental health were drawn largely from these studies, which were supplemented by later interviews and the conferences cited at the beginning of this chapter.

To initiate the studies, the sponsors brought together for a two-day workshop at the Rosenstiel School more than 40 knowledgeable participants, the majority of them scientists, leavened with a seasoning of international experts, among them policy specialists, conservationists, international civil servants, and experts on either the Caribbean or Latin America.

Most were from the United States, though several hailed from each of the regions under consideration. The study can be faulted for that, though as a pilot project, it did probably did not matter. To get ahead of the story, however, the participants judged that, in proceeding with further case studies, it was vital to assure that such studies had a truly international flavor.

The studies tested an operational concept, described below, for environmental security. Science-based case studies that mesh physical science with political economy are suitable vehicles for forecasting conflicts and lesser tensions, derived in varying degrees, from environmental stress.

The studies tested an operational concept, described below, for environmental security.

Science-based case studies that mesh physical science with political economy are suitable vehicles for forecasting conflicts and lesser tensions, derived in varying degrees, from environmental stress.

Case studies are not new, dating back at least to Thycidides. With respect to environmental security, they have been used—with more or less rigor—to show historical linkages between environmental insults and particular conflicts, for example in the case of Haiti, discussed earlier here.

The novelty of these studies lies in their prospective nature and inherent replicability, in seeking to forecast the risk of conflict (or conflictive relations short of war) derived from changes in the environment. In an experimental way, they joined marine and atmospheric science with political and societal analysis, to suggest where ecosystem change may stimulate conflict.

The choice of dissimilar ecosystems was deliberate, to test whether the North-South Center/Rosenstiel School case study framework might have general applicability. Not suprisingly, the security profiles of the two regions differed markedly.

By design, the studies were analytic and did not prescribe policy. They were intended, however, to help illumine policy choices. Whatever their defects, among them an absence of well-defined thresholds to identify where ecosystem change crosses the line to become a security risk, the studies support two judgments:

1. Environmental security has practical value. In the environmentalists' metaphor, the concept is a "canary in a coal mine," a warning of conflict or signficant tensions, with nations and across borders.
2. The impact of environmental stress is addressed in concert with political, economic, and social factors. By relating physical data to their potential consequences, science-based case studies can provide critical insights.

Method

Using a tested environmental risk model from the Environmental Protection Agency (EPA), the participants broke into working groups for the two cases, to conduct "brainstorming" exercises with respect to environmental trends over the next ten years. Each group tested its findings on the other and then refined them. The next day, the groups sought to translate specific environmental trends into economic, social, and political consequences, and to their forecast implications for conflict or significant tensions.

They eschewed predictions, which imply certainty in favor of forecasts of alternative futures. By combining a tested risk model with corporate forecasting techniques, notably alternative futures, and relying on a mix of tools, scientific data, indicators, and classical analysis, the studies arrived at largely consensual forecasts.

Guidelines

In preparing the studies, the participants put together six working guidelines, which may be worth bearing in mind for future work.

1. Environmental security is a subset, not only of security in the traditional sense, but also of another great concern: sustainable development, defined as the wise use of the environment in the service of sustainable, long-term growth.
2. Environmental change may help spark violence, but more frequently (or in early stages) it leads to conflictive relations, within or among states, that

take a toll on the workings of economies, the effectiveness of political institutions, and the coherence of societies.

3. The most promising method of analysis is to begin with scientific data, incorporating as the analysis proceeds, economic and political variables. The end products are forecasts, no straight-line prediction, which seek to bound the possibilities for conflict in geographically defined situations.
4. In examining prospects for internal and cross-border tensions, keep in mind that environmental security may be a two-way street, involving a reciprocal relationship between conflict and enviromental damage.
5. It is essential to choose a manageable time frame—no more than ten years. To the extent possible, case studies should be framed to tell a compelling story to the nonscientist. Without a reasonable time frame and a persuasive story, such studies, however valid, are unlikely to receive attention.
6. Territorial disputes over maritime, riverine, and land boundaries, have always been indicators of tension, "canaries in a coal mine," as it were. More and more, their analysis requires attention to environmenal factors.

The point in describing the methodology for these studies before proceeding to the conclusions lies in their potential replicability. The method needs refinement, as the participants themselves pointed out, but it has predictive value, judgments reinforced at a subsequent government sponsored joint U.S.-U.K. conference about environmental security, which reviewed these studies.

Sources of Conflict

A principal task for the case study was to identify potential sources of conflict that might arise from environmental vulnerabilities. Grouped under five headings for purposes of analysis, each of these factors may also be viewed as a prime indicator of possible conflicts: (1) disruption of economic structure, (2) resource degradation leading to more resource competition, (3) human health, (4) territorial disputes, and (5) policy management.

For example, environmental change is spurring a resurgence of malaria and transmission of dengue, cholera, and other intestinal illnesses. Will diseases stimulate conflict? Not by themselves, but they levy higher costs on fragile economies, stress health systems, and lessen productivity. Similarly, while environmental changes have not caused territorial disputes (such as those between Nicaragua and Costa Rica, Haiti and the Dominican Republic, Nicaragua and Honduras, Trinidad and Tobago and Venezuela, and Panama and Colombia), these changes appear to be worsening the disputes. A cycle may be under way in which disputes hamper environmental management across land and maritime frontiers, and outcome that, in turn, further aggravates tensions. And an absence of coherent national policies and international arrangements for wise use of environmental assets makes it difficult to achieve sustainable growth, manage resources, and provide accountability for sources of pollution.

In the best of circumstances, over the next five to ten years environmental degradation will increase in the region. But the extent of that increase will be contingent upon how the nations of the region and the international community deal with the challenge. This finding led the study to forecast two broad alternative scenarios.

- *Lower Probability*: More effective national and international strategies, underpinned by science and hard data, to contain environmental loss and limit its economic and political costs. Growing cooperation among the Small Island Developing States and within Central America, extending to the United States and other Organization for Economic Cooperation and Development (OECD) nations. *Illustrative measures*: International arrangements to restore the Caribbean fishery, mediation/arbitration of territorial disputes and revitalization of cross-border conservation, steps to mitigate future disasters, and the achievement of sustainable growth in a majority of the states.
- *Higher Probability*: Business as usual, largely unchecked environmental damage, leading to spreading economic losses and, in some nations, political unrest and unhinged democratic institutions. Criminality and recurring flows of economic refugees from the Wider Caribbean and the rest of Latin America. Territorial disputes undermine international cooperation on various fronts and, in some locales, prompt low-level violence, perhaps involving militaries.[3]

Major Conclusions

- In the Wider Caribbean, environmental stress will contribute to conflictive relations, not only within countries, but also among them, examples, Haiti and the Dominican Republic, Nicaragua and Costa Rica, Nicaragua and Honduras, and Colombia, Panama, and Colombia, Trinidad and Venezuela, and intra-Caribbean disputes over ocean and coastal pollution, fisheries, migration, and trade.
- In the next five to ten years, no serious conflict is likely to break out in the UPB/Pantanal. Much is at stake, however, for Mercosur partners. Intra-regional tensions over the environment and the Hydrovia (projected to be a much larger version of the channelizing of Florida's Kissimmee River) are

likely. These tensions, involving winners and losers among local stakeholders, are likely to shake regional and national politics.

- No single country can manage the environmental stresses affecting economies and societies of these two regions. More effective international cooperation is required, along with the further development of legal and institutional frameworks for their management.
- U.S. policy faces an immediate environmental challenge in the nearby Caribbean. Not so for the distant Pantanal, where Brazil is the giant among the players. However, the Pantanal is an important test case for devising and implementing international legal regimes and institutional arrangements for managing environmental assets.

Notes

1. Haiti ranks among the most studied examples of environmental security; particularly noteworthy is Thomas Homer-Dixon's analysis of the relationship of forest scarcity to poverty and violence. The official view—that of the international community, including the United States—of the roots of conflict coincides with the academics who judge deforestation to be a major cause of Haiti's tragedy.
2. Deudney, *The Case Against Linking Environmental Degradation*.
3. Frank McNeil and Jeffrey Stark, "Thinking about Environment Security: Southeast Asia and the Americas in Comparative Perspective," Working Paper Series (Miami: The Dante B. Fascell North-South Center, October 2001), 17.

CHAPTER SIX

Environmental Challenges in the Panama Canal Watershed

Darci Glass-Royal and Ray C. Simmons

Environmental issues increasingly shape economic development, urbanization and migration patterns, infectious disease outbreaks, natural disaster vulnerability, and regional cooperation and stability in ways that bear significantly on U.S. national interests. These environmental stresses can significantly undermine regional political, economic, and social stability and require U.S. financial, diplomatic, and/or military intervention.

For the U.S. national security community to identify opportunities to mitigate environmental problems before they become regional crises, it is essential to understand the impact of these issues across political, economic, and social sectors and their subsequent impact on regional dynamics. Effective analyses of environmental issues must, therefore, consider the requirements and perspectives of military organizations, civil and international government agencies, industry, and academia. Further, these analyses must provide evaluations to enhance support to the development of coordinated, interagency strategies for planning and engagement consistent with national and regional security objectives. This chapter provides a case example of how environmental degradation in the Panama Canal Watershed, if left unchecked, could undermine national economic stability with potential attendant consequences for wider national and regional instability.

Background

December 1999 marked the official transition of the Panama Canal from an entity operated by the Panama Canal Commission, a U.S. Federal Government agency, to one fully owned and operated by the Government of Panama. This turnover raised concerns, both within the United States and among the international community, regarding the continued security and viability of the Canal. Concerns ranged from technical issues regarding Canal operations to labor unrest to foreign control of the Canal. Ironically, the problem most likely to threaten the future of the Canal—diminished water resources—received only minimal attention at best.

In congressional testimony in October 1999, Commander in Chief, U.S. Southern Command, General Charles Wilhelm testified that environmental degradation and watershed mismanagement leading to increased siltation is a significant threat to the future of Canal operations. In recognizing the issue of environmental degradation as a critical linchpin affecting the future availability of water to operate the Canal, General Wilhelm focused attention on a problem that has long plagued the Panama Canal Watershed (PCW). Specifically, land-use patterns in the watershed have resulted in the clearing of large tracts of primary forest and accelerated both soil erosion and the runoff of human, animal, agricultural, and industrial waste into the rivers and lakes that provide the fresh water necessary to move ships through the Canal.

As container ships continue to increase in size, thus exceeding the current dimensions of the Panama Canal, and as shipping routes increasingly rely on cross-country rail alternatives, the question inevitably arises as to the continued cost effectiveness of the Canal. According to maritime industry analysts, many shipping lines are opting for hub ports on each coast for goods destined for the United States, combined with cross-country rail routes specifically to avoid the issue of ship size restrictions in the Panama Canal. The fact remains, however, that the Panama Canal is a critical node in the movement of essential products worldwide (particularly for the Caribbean and the Pacific side of Latin America) and any significant degradation of Canal operations will create serious economic and political repercussions well beyond Panama's borders.

Environmental Impacts on Watersheds

A watershed is the area drained by a river or river system. Under ideal conditions, watersheds provide sustainable water resources to meet the demands of domestic consumption, industry, agriculture, and farming. The extent to which any watershed can meet these demands depends on balancing water supply and water demand.

As populations grow, water demand increases, placing more stress on the supply side of the equation. However, even in situations where demand is constant, the water supply can be significantly decreased by environmental factors that degrade the ability of the watershed to continue providing either the quality or quantity of water required. In cases where supply and demand are closely equilibrated, any environmental factors that degrade the watershed can significantly disrupt water availability to one or more sectors. Therefore, managing the interrelationship between supply and demand (particularly when there are frequent shifts on the demand side) is essential to effective watershed management and policy decision making.

Climate

Panama, like other tropical countries, is subject to short and long climate cycles. The short cycles are characterized by annual wet and dry seasons associated with the seasonal meandering of the Inter-tropical Convergence Zone (ITCZ) typical of tropical climates. The long cycles are characterized by multiple years of long, high rainfall rainy seasons followed by multiple years of short, low rainfall rainy seasons. In order to stabilize water availability throughout the long cycles, a storage reservoir must be created as a buffer to mitigate the effects of prolonged dry periods associated with the long cycles.

In addition, the El Niño/Southern Oscillation (ENSO) effect is intermittently interspersed within the long cycle. ENSO events can create periods of prolonged drought with significant reductions in water table levels. An ENSO event in 1997 reduced the water level of Lake Gatun sufficiently to require the imposition of draft restrictions on ships transiting the Canal from 39.5 feet to 37 feet. The rhetoric surrounding this event was substantial, and many Panamanians believed that Panama came close to having to choose between water for ships and water for people. Others, however, argue that the storage capacity of Lake Gatun

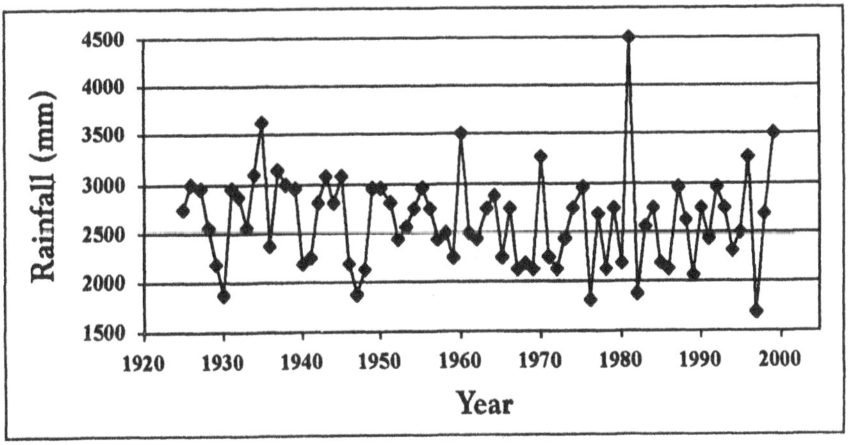

Figure 6.1. Panama Precipitation Graph

was fully adequate to provide a buffer for this dry period, and that it was mismanagement, related to release of water from Lake Gatun in advance of a large storm that was anticipated but that never materialized, that created the crisis. In either case, there is no doubt that in order to stabilize water availability and ensure operation of the Canal at optimal capacity, Panama will have to develop water storage capacities in the future that exceed those provided by Lake Gatun.

Watersheds

Precipitation that does not evaporate into the atmosphere is transported through the watershed as either surface water or groundwater. Groundwater flows underground and, depending upon a number of factors—the type, amount, and duration of precipitation; topography; soil type; land use—eventually emerges as surface water, a process that can take anywhere from a few hours to decades.

A healthy watershed can provide a sustainable water source for generations. However, the complex system of water flow in a watershed is highly susceptible to outside interference. Deforestation, agriculture, road construction, and improper sanitation are among the greatest threats to maintaining optimal water capacity in a watershed. This human involvement need not, and, in fact, often can or should not be eliminated, but regulation based on a clear scientific understanding of the particular watershed environment is essential to ensuring water sustainability.

Deforestation

The effects of deforestation on a watershed's ability to maintain optimal water capacity are both complex and controversial. Deforestation decreases water storage capacity and increases runoff, which is particularly threatening during the dry season, when the soil's capacity to absorb water is diminished. Further, forest canopy and ground litter (accumulated, compacted leaves and other debris) play a significant role in preventing siltation due to erosion.

Most scientists acknowledge a relationship between deforestation and areal precipitation rates, though the concept and theories remain controversial. Amounts and frequency of rainfall are dependent on several factors, including geographic location, topographic features, and sea-surface and land temperature. Vegetation cover also appears to have an effect, though, again, it is much debated within the science community as to the extent and magnitude of its direct role.

Reforestation of previously deforested land may not be the optimal solution to increase water retention. The litter created on the forest floor is actually more essential than the trees themselves for absorbing rainfall. For this reason, native rain forest trees are not the only plants effective in reducing moisture loss. The introduction of grasslands or cash-generating trees, e.g., citrus orchards, which produce profit and increase local jobs, is generally much more cost-effective than reforestation with native trees, and often is, in fact, equally adequate in retaining moisture and limiting erosion. In Panama, this is especially relevant, as reforestation efforts there are of significantly higher cost than anywhere else in Central America, due principally to the U.S. dollar-based economy. In regions not completely deforested, existing forests can be protected by selected harvesting techniques which include cutting trees based on species composition, age classes, and necessary spacing between cuts, to allow for optimal regeneration.

Agriculture

Improper agricultural techniques in deforested regions are more devastating to the watershed than the deforestation itself, with erosion and toxic runoff, the greatest threats. Along with sedimentation from erosion, toxic runoff from agricultural chemicals and livestock waste invades the watershed, polluting both groundwater and surface water. Nitrogen and phosphorous, both found in animal waste, though necessary to maintain aquatic life, become toxic in large amounts. This "nutrient pollution" causes algae blooms, which both destroy vegetation and aquatic animals, and make water impotable. In the case of Panama, algae blooms in the Canal would severely inhibit ship traffic due to mechanical fouling.

Urbanization

Urbanization is a significant threat to any watershed. Road construction is of particular concern. During construction, soil displacement increases erosion and sedimentation. After the roads are built, solid (concrete, asphalt) surfaces inhibit water retention and increase water runoff. Further, like agricultural pesticides, exhaust from automobiles contains harmful chemicals that contribute to toxic water levels. These effects can be significant, even where deforestation has not occurred. In the case of Panama, the Boqueron region, perhaps the most highly forested region of the watershed remaining, has the highest siltation rate due principally to road structures built for mining operations. In addition to road construction, improper sanitation practices as populations increase in the watershed, along with waste from established urban settlements and industrial sites, add toxic and nutrient pollution to both groundwater and surface water systems.

While these activities cannot be eliminated, maintaining a buffer of forest land between high-impact activities and water sources can mitigate the effects of pollution. Forests surrounding water sources have the potential to trap sediments, remove pollutants, and take up nutrients such as nitrogen and phosphorus from surface runoff and shallow groundwater. In addition, careful planning to keep the most sensitive watershed areas free from high-impact activities increases the likelihood of clean, sustainable water sources.

Institutional and Political Issues

When the Panama Canal Authority (PCA) assumed full responsibility for operation of the Canal, it assumed not only the responsibility to operate a system of locks, but also the responsibility to serve as caretaker for the rivers and waterways that are the source of fresh water for the Canal. With 52 million gallons of fresh water lost during each ship transit, ensuring a sustainable, on-demand supply of fresh water for Canal operations is essential to the continued viability of the Canal. The challenge for the PCA, then, will be to transition from an organization that manages a canal to one that manages a watershed, as well.

Under the best of circumstances, managing a watershed is a complex undertaking requiring an in-depth technical understanding of climate cycles, forestry and natural resource management, transportation planning, demographic patterns, and community relations. The dual factors of a new presidential administration and a newly formed Canal organization with newly acquired autonomy and an expanded sphere of responsibility create a significant challenge for the Government of Panama.

Law 19, passed in June 1997, established the Panama Canal Authority as an autonomous agency of the Government of Panama responsible for operating, managing, and maintaining the Canal. An administrator and deputy administrator, under the supervision of an eleven-member board of directors, manage the PCA. The administrator serves a seven-year term. Ten of the board directors are appointed for a nine-year term by the president of Panama, with consent of the cabinet council, including one member selected to serve as minister of state for canal affairs. The eleventh board member is chosen by the legislative branch and may be freely appointed or removed. Under Panama's Organic Law (an amendment to the constitution defining the legal position of the PCA) the PCA is granted final authority over all projects within the Panama Canal Watershed.

Law 21 is the primary legislation governing land use in the watershed. Although Law 21 was based on a costly, multiyear study, the law has been extremely controversial and has been criticized for enacting an "architect's view of land use" that bears little relation to the actual productive capacity of agricultural land and that fails to consider the rural economy or the requirements of the various segments of society living in the watershed.

Law 44, passed in August 1999, established the expanded watershed area and annexed 213,000 additional hectares to the original watershed. When developed, the additional watershed area will provide about six times the water storage capacity of the original watershed and could easily double the water resources available for Canal operations.

Although Laws 19, 21, and 44 establish the basic parameters governing the management of the Canal and the watershed, these legislative documents only begin to depict the complex political, technical, and bureaucratic issues that bear on environmental issues as they impact Canal operations. With respect to the PCA, several important factors must be kept in mind. The PCA is *not* a replacement for the Panama Canal Commission (PCC), the U.S. federal agency responsible for the Canal prior to the turnover in December 1999.

The PCC, as an arm of the U.S. government, had immediate access to multiple sources of information and expertise for environmental and scientific data through organizations such as the National Oceanographic and Atmospheric Administration and the U.S. Forestry Service. The PCC operated on a cost basis and had little or no need to rely on the international donor community to fund operations or provide technical assistance. In contrast, the PCA has no formal access to these sources of information, operates on a more limited budget, and will need to rely on both the foreign donor and development banking communities while lacking experience collaborating with these organizations.

Under Panamanian law, the PCA is responsible for providing water in the watershed for municipal and industrial (M&I) consumption, Canal operations, and hydroelectric power. This charter would imply that the PCA also exercises authority over the water distribution system. However, this is not the case, and while reliable figures estimate that 50 percent of water is lost in the distribution system due to aging infrastructure (i.e., leaky pipes), the PCA has neither the inclination, authority, budget, nor the responsibility to upgrade the water delivery system.

Further, rising population rates are continuing to add M&I demand for water. Yet, there are virtually no incentives in place to encourage conservation because the Government of Panama maintains a water pricing structure that charges unreasonably low rates for water relative to actual operating costs and water availability. The current water pricing structure was established in a period when there was an abundance of water relative to M&I demand. Despite the need to increase water costs to municipal users, in particular, any suggestion of raising rates is politically charged. Increasing water rates is sensitive because it is not consonant with the populist platform of President Mireya Moscoso and because the PCA already is perceived as wielding too much power in the government structure.

The tension over the disconnect between the organization responsible for managing the water (the PCA) and the organization responsible for delivering the water (IDAAN, Instituto de Acueductos y Alacantarillades Nacionales) is exacerbated by the PCA's somewhat strained relations both with other Panamanian Government organizations and Non-Governmental Organizations (NGOs). The PCA is granted absolute authority for projects within the watershed, but when the watershed area was expanded, this brought the PCA into direct opposition with other groups that had formerly been granted jurisdiction in these areas.

The PCA seems to lack a broad vision of how it integrates with the rest of the government and, in many cases, its decision-making capacity directly conflicts with the mission or authority of organizations such as ANAM, Panama's environmental authority. The ability (and plans) of the PCA to undertake development projects within the watershed directly pits the PCA against international or NGO groups working to preserve protected biological corridors that, in many cases, include lands that fall within the original or expanded watershed.

The CICH (Comision Inter-institutional de Cuencas y Hoyas), an intergovernment coordinating committee, has been mandated to facilitate coordination between the PCA and the rest of the Panamanian Government. While a good idea in theory, the CICH in reality is a highly political organization chaired and controlled by the PCA. The PCA selects the organizational members and has the right to exclude key government organizations. Two NGOs also are included as CICH members. However, it has been privately acknowledged that the member NGOs were selected as a "preemptive strike" to engage the Catholic Church, in particular, as a means of co-opting local populations living in the expanded watershed areas where the PCA anticipates undertaking new development projects.

The PCA is not unaware of the sensitivities surrounding their elevated status in Panama. PCA representatives publicly state that they will serve in a "coordinating," not an "executing" function, and discuss development projects in the expanded watershed in terms of them being *national* projects to support the water needs of the country as a whole rather than projects designed to meet the water needs of the Canal only.

In general, however, the PCA lacks experience in dealing with the broad array of environmental and social issues that are inherently tied to any further development of projects intended to increase water resources for the Canal. The PCA maintains an environmental compliance division, but it is staffed by technical experts who, by their own admission, lack experience with land management and sustainable environmental development. The PCA is working with the Inter-American Development Bank for pre-feasibility studies related to development projects in the expanded watershed, focusing in particular on strategies for dealing with rural communities. The perception among other development banks and the NGOs, however, is that these efforts are more public relations campaigns than substantive efforts to mitigate environmental or social issues.

Demographic Issues

Population growth and migration are inextricably linked to threats to the Canal watershed and ultimately the Panama Canal itself. The population of Panama is approximately 2.8 million (mid-2000) and is projected to double by 2030. More than half of the population is concentrated in the former Canal Zone and Panama City, with the remainder residing primarily in the Pacific lowlands west of the Canal.

The largest population center is in and around the capital city of Panama City, with an estimated population of well over one million people. While Panama's overall population growth rate of 1.7 percent is one of the lowest in Central America, Panama City's rate of growth is 4.3 percent, placing an enormous strain on municipal services.

Population has steadily grown in the rural areas of the original watershed since the 1960s and is now estimated at 150,000. This growth has been spurred, in part, by road improvements and development in areas that were once inaccessible jungle. Estimates of population in the area of the expanded watershed range anywhere from 8,000 to 30,000, primarily indigenous peoples. Proposed development projects in the expanded watershed are likely to precipitate a wave of migration into the cities and the original watershed, further stressing water availability and delivery systems.

Panama's economy is dominated by the service industry. This sector employs one half of the population and accounts for two-thirds of the gross national product (GNP). Agriculture on the other hand employs approximately one-third of the population and accounts for one-tenth of the GNP. The economy of the people residing in the watershed is mostly rural, with subsistence farming as their only means of survival. Poor farming practices in the watershed are becoming an increasing threat to the watershed through increased deforestation and subsequent increased runoff.

The Original Watershed

The original PCW spans an area of 326,225 hectares and encompasses the Chagres River basin, including Lakes Alajuela and Gatun, as well as the Indio, Caño Sucio, and Coclé del Norte River basins west of Lake Gatun in the Atlantic region. The fifty-one-mile-long Canal channel traverses the Isthmus of Panama northwest to southeast from Colón on the Caribbean side to Panama City on the Pacific side. Three man-made lakes, Gatun, Alajuela, and Miraflores, provide the controlled water supply for the Canal. Lake Gatun, on the Atlantic side, is the largest of these lakes, encompassing 165 square miles. Lake Alajuela, to the east of Lake Gatun, serves as a supplemental reservoir to maintain water levels for the Canal during dry periods. In addition to providing all the water for the Canal, the PCW also is the principal water source for numerous industries within the watershed and for most domestic consumption in the cities of Colón, Panama City, and Arraiján.

Law 21 governs the specific use of land in the watershed. The law, based on a $6 million, eighteen-month study, is now considered by many to be seriously flawed, as it is generally recognized that the underlying study failed to take into account hydrology or the impact the law would have on the existing population. Several international groups have voiced concern over the current treatment of the watershed. The Smithsonian Tropical Research Institute (STRI) has been warning for years that population pressures, corruption, and unchecked urbanization will irretrievably damage the watershed and ultimately have a detrimental effect on Canal operations.

Water supplies and, hence, precipitation, sedimentation, and siltation rates, are the greatest concern within the PCW, and while there is no dispute that significant deforestation has taken place in the watershed since 1952, there is much disagreement over the exact effect of the deforestation on the watershed. On average,

Panama has cleared its forests faster than any other Central American country—a rate of 148,000 acres per year.

By 1978, deforestation of almost the entire watershed was complete in all but the immediate Canal Zone. Notably, however, precipitation-monitoring sites throughout the country have not indicated any appreciable change in precipitation over that time period. Sedimentation/siltation rates, however, have consistently been a problem—by 1994 approximately 8 percent of Lake Alajuela's storage capacity had been reduced due to sedimentation, with an additional 6 percent reduction anticipated by the end of 2000. Most experts are reluctant to recommend reforesting areas that have been cleared, except along stream banks, and generally agree that planting grasses in the deforested areas would be just as effective in slowing runoff and erosion as trees.

While forestlands are officially under the control of the Panamanian government, rules requiring permits to clear forests are seldom enforced. As a result, farmers, loggers, and industry clear approximately 2,000 acres of forest each week. Halting this activity would not be an easy task. Local farmers need the land to grow crops and raise livestock in order to survive. Many farmers acknowledge that their slash-and-burn techniques are detrimental to the watershed, and know that the local drinking water is threatened, but as one farmer stated, "Why should I care about the Canal or the people of Panama City if it will mean my family can't eat?"

The ramifications of lost forestland in terms of climate, weather, and erosion are not fully understood, and the PCA views deforestation solely in terms of *trees*. Hence, since precipitation rates do not appear to have changed appreciably since the 1950s (before much deforestation had occurred) and many scientists believe that vegetation in general (particularly grasses), is as effective as trees in controlling erosion, the PCA does not currently consider deforestation a direct significant threat to the PCW. However, as approximately 10 percent of the western watershed forest disappears every year, without some form of erosion control, a significant threat to the watershed and the Canal will exist in the future.

Water Demand Projections

Water requirements in Panama are defined in terms of "lockages," an economic unit derived from the costs for a ship to transit the Canal, but not directly translatable into precise water amounts. The current PCW is capable of supplying a nominal thirty-nine lockages per day, while current requirements are: Canal operations, thirty-four lockages per day; and M&I, four lockages per day. Upon completion of the Gaillard Cut widening (circa 2001), the additional six ship transits per day will require an additional five lockages per day for full, unrestricted forty-three ships per day operations, increasing *Canal requirements alone* to thirty-nine lockages per day. PCA officials have expressed confidence that during wet periods the PCW will be able to meet this demand; however, during dry periods, they acknowledge that the PCW will *not* be able to meet minimum requirements and that Canal ship traffic will have to be strictly controlled, either by reducing daily transits or imposing draft restrictions (as during the 1997 ENSO event). Regardless, it is estimated that through population growth alone, sometime in the 2010–2020 time frame there will not be sufficient water within the current watershed to meet M&I demand and operate the Canal within tolerant (for the shipping industry) reliability standards. Further, Panama City's population is expected to double by 2030 and current projections, based on the summer 2000 study, estimate that lockages required for M&I alone will double (to eight) by 2050.

Further, maintenance of the nearly one-hundred-year-old locks is also a significant issue. It is estimated that proper maintenance of the locks will, in the near future, require a rehabilitation effort of twelve to eighteen months. To achieve this under the current two-lock system, the Canal would be restricted to single lock operations, which would undoubtedly lead to significant transit backups that easily could double the current twenty-four-hour maximum transit (including wait time) standard.

Both these issues impact Canal reliability. The ability of shipping companies to accurately predict transit times, port dates, and costs is a critical factor in selecting a shipping route, and potential degradation in Canal efficiency and capacity is a major concern for the PCA. From a study completed in the summer of 2000, PCA officials estimate that Canal reliability could fall during excessive dry periods to as low as 52 percent, exceeding ship industry tolerance levels. This is a significant problem, as the PCA believes that it cannot afford to lose *any* ship traffic to other sources, because the market is becoming increasingly competitive and once shipping lines choose alternative routes or hubs, they are not likely to return to using the Canal, even if reliability could be restored. Traffic lost once is thought to be lost forever. Further, it is not only Canal revenue that would suffer, as a loss of toll revenue translates into at least a double or triple loss of *national* revenue because of the loss of collateral business (e.g., shipyard, banking), as well. A number of alternatives to address this concern have been suggested; foremost, expanding the current watershed to create additional water storage capacity and building a third set of locks to increase ship traffic capacity.

Building a third set of locks is a controversial issue. The expense alone, estimated at $5–7 billion, exceeds the cost of the original Canal project (using current year dollars). Proponents argue that an additional 25–40 percent more traffic could be handled in the Canal by a third set of locks. The PCA proposes offsetting the cost with the correspondingly greater revenues generated by the increased capacity and by selling hydroelectric power that would be generated by the additional dams necessary to provide the additional water that would be needed. Outside experts question whether the market exists within the region for the amount of hydropower that would be necessary to offset the substantial cost of the new locks. In April 2000 the Government of Panama announced its intent to build the third set of locks, but that announcement was rescinded in July 2000, and the current status remains under debate.

Beyond growing demand from a burgeoning urban population, poor infrastructure and institutional attitudes held over from the 1970s, when the United States overproduced water and electricity, are exacerbating many of the water shortage problems facing the watershed in the future. Currently, 80 percent of Panama City's residents rely on the watershed for clean drinking water. Of the four lockages currently allocated for M&I use, as much as 50 percent is lost due to aged and broken underground pipes, and more is lost to waste. Simply stated, even if no one lived outside the cities of the original watershed, if urban population growth continues at the current rate of 4 percent, if no action is taken to remedy the flaws in current water distribution infrastructure and policy, or to increase the amount of water available, there will not be enough water for M&I requirements and reliable Canal operations by 2020.

The Expanded Watershed

Though there clearly are significant efficiencies that could be gained within the existing watershed by repairing/upgrading water delivery infrastructure, correcting the water pricing structure, promoting conservation, and minimizing sedimentation, there are compelling reasons for pursuing water development projects outside the current PCW. The annexation of 213,000 additional hectares to the PCW under the guidelines of Law 44 is the first step in the PCA's quest to increase the number of lockages that will be required to ensure a sustainable source of water for increased M&I demand and to meet Canal operating requirements in the future.

If fully implemented, the expanded watershed could increase the storage capacity of the original watershed by a factor of six, significantly increasing the amount of water available for Canal operations and M&I. This increased capacity would ensure sufficient water to meet the M&I demands of the growing population for the next one hundred years, while preserving the ability to operate the proposed third set of locks should it eventually be decided that the locks will be built. Further, infrastructure developments in these remote areas could bring much needed services and job opportunities to a generally impoverished region and perhaps entice people to either stay or migrate there, relieving some of the growth pressures on the already stressed urban centers, particularly Panama City.

In considering alternatives to increase and stabilize water resources for the Canal and M&I use, the PCA claims to have reviewed virtually every option and potential location. The western watershed project was assessed to be the only truly viable solution. However, there are numerous social, economic, and environmental concerns that the proposed project raises, not the least of which is cost. Estimates place the cost at $15 billion, which will clearly require foreign/donor assistance. While the PCA claims that hydropower generated from the new dams proposed for the expanded watershed project will offset costs, economists believe it unlikely that there is a regional market for that much hydropower. Further, political tensions with displaced local populations (many of whom have lived in those regions for decades or even centuries), the Catholic Church, and other humanitarian assistance support organizations and environmental/conservation groups will be very difficult.

Although a number of (uncoordinated and expensive) studies are under way, no good assessment of the rural economy or accurate census of the population in the proposed expanded watershed exists. Official population estimates for the region range from as few as 8,000 to as high as 35,000. In fact, no one knows for sure, but there is general agreement that the proposed expansion program, when implemented, will likely spur a wave of migration into the cities, exacerbating the very problem of growing urban M&I demand that the project is designed, in part, to alleviate.

Legitimate concerns have been expressed regarding the tremendous cost of the proposed development of the expanded watershed, with no clear financing or possibility for recapitalization of the investment in sight. In particular, critics of the plan argue that the hydropower plan, proposed to offset the cost, is unrealistic, particularly because power currently produced by hydropower projects is being sold to the city of Colón at rates substantially less than fair market value. Further, plans for the expanded watershed presume significant projections of increased Canal traffic (nearly double) and collateral revenue growth (e.g., banking, insurance, maritime services) which may prove overly ambitious.

While efforts to market the plan as a "national" project rather than a "PCA/Canal" project helped to ensure passage of the law, the lack of concrete public information has spawned rumors which have led to the formation of civic groups within rural communities who oppose construction. The Catholic Diocese of Colón, which has considerable influence in the area, considers the watershed expansion a human rights issue, insisting that government estimates of indigenous populations that the venture will displace is only about half the actual number of people that will be left homeless after development is completed.

In light of overwhelming evidence that future water shortages for the Canal and population at large are inevitable, the Government of Panama and the PCA, in particular, clearly face significant challenges. While taking steps to repair/upgrade existing infrastructure systems and policies would go a long way toward alleviating shortages in the near term, expansion of the current watershed appears to be the only viable long-term solution. Yet, expansion plans are wrought with economic, political, and social perils. A clear, well-defined development plan for expansion, based on sound scientific, technical, and sustainable principles will be essential to ensure sufficient, clean water will be available to meet the requirements of both the Canal and M&I over the next fifty years.

What Can Be Done: Conclusions and Options

Given the complex interplay of environmental (and economic, social, and political) factors in the Panama Canal Watershed, as well as the significant uncertainties regarding key science issues, it is difficult to sort out what should be done from what can be done. In the words of the PCA, "the western watershed development projects are an engineer's dream of a lifetime," but technical feasibility alone is not sufficient justification for moving ahead with major watershed development activities.

Five fundamental points should drive policy options at the most basic level:

1. While there is sufficient water (thirty-nine lockages) in the original watershed to support current M&I demand of four lockages per day plus thirty-seven ship transits per day, upon completion of the Gaillard Cut expansion, there is *not* sufficient water to reliably (by shipping industry standards) support current M&I and forty-three ship transits/day.
2. By 2010, under current population growth estimates and with forty-three ship transits per day, an additional six lockages per day will be required to meet minimum water demand.
3. Sedimentation, distribution losses, and waste are reducing water capacity in the original watershed.
4. Erosion control, conservation measures, improvements in water delivery systems, and restructuring of the water pricing system would increase water capacity in the original watershed by five to seven lockages per day, sufficient to support the likely 2010 M&I demand.
5. Even with gains realized from these water saving measures, without watershed expansion beyond 2020, there will not be sufficient water to meet the increased M&I demand associated with the growing population and ensure reliable Canal operations.

Certainly, reducing ship transits per day would free up water for M&I demand and this, of course, is ultimately the issue at the heart of managing the PCW. At what reliability level must the Canal operate to ensure shipping interest confidence, and hence, regional and global access to critical goods? What level of toll and ancillary revenue is essential to Panama's stability? And how much tolerance is there within Panama for degraded M&I access to water in order to maximize ship and collateral revenues? Thus, it becomes clear that managing environmental threats to the PCW in order to maximize water availability bears directly on issues of national and regional stability.

In the near term, every effort should be made to increase efficiency, reduce waste, and prevent further degradation of the original watershed. These measures might include:

- establishing water use fees in line with delivery costs and available supplies;
- fostering public education programs to encourage conservation;
- improving environmental standards and enforcement for roads, housing, and sanitation, particularly in the former Canal Zone where housing developments are flourishing, as ARI sells off the reverted property to private ownership;
- encouraging technical assistance by outside industries and governments;
- reducing sedimentation and runoff through sound environmental development practices (including paving Pipeline Road to a fifty-year standard, managing runoff from chicken factories near Lake Alajuela, and reestablishing vegetation, especially with grasses and citrus);
- repairing the aging water delivery infrastructure; and
- directing a small percentage of Canal toll revenues (<1 percent) and developing institutional mechanisms for collecting fair use fees for parks and lakes that are set aside for environmental management of the watershed.

In the longer term, undertaking some level of development in the expanded watershed area will be the only solution to increasing water availability. Which projects, at what cost, and when will be the key issues that must be resolved. The economic burden of these projects will be substantial, as will the need for technical assistance. Perhaps the most important step that can be undertaken at this point is the complex task (given the large area and remote location) of gathering data to develop a sound, fundamental environmental baseline for the expanded watershed area, from which responsible, fact-based technical and policy decisions can be confidently made.

Finally, the issue of environmental degradation in the Canal watershed has been largely unrecognized and unacknowledged as a serious threat to the future of the Canal. However, the effects of environmental degradation on the future availability of water to support Panama Canal operations are more fundamental to the broad underpinnings of canal security than any of the issues that were flagged as potential threats to the Canal. Canal revenues and collateral maritime activities are central to the overall economic stability of Panama. Any significant decline in the vitality of the Canal, including a decline precipitated by environmental degradation in the watershed, could generate instability in Panama with a very real potential to spill over into the broader hemispheric region. Environmental degradation in the Panama Canal Watershed, therefore, is directly linked to both Panamanian and regional security and stability.

CHAPTER SEVEN

Water: The Hydraulic Parameter of Conflict in the Jordan River Basin

Stephen D. Kiser

"Unless properly managed, water scarcity can be a major source of strife, as well as a roadblock to economic and social progress."[1] These words, spoken by former U.S. secretary of state Madeleine Albright, illustrate the expanding notion of security. Instead of viewing it through the narrow lens of traditional national security, the less discriminating notion of human security is gaining a broader acceptance in the international community. In a world full of conflict, identification of the causes of fighting is as important as the identification of who is fighting, and the conventional premises of national security are no longer sufficient. Environmental security is one key component of human security, and access to water is one of the components of environmental security—one that is increasingly acting as a catalyst for different groups to initiate hostilities, or at least as a hindrance to achieving peace and stability.

The Middle East and the Jordan River Basin highlight this trend well. Despite the spiraling chaos and near daily escalations between Israelis and Palestinians that seemingly makes peace an increasingly remote prospect, most major parties involved at least appear to eventually want peace. In a region of the world where ten years ago simply admitting your enemy even had the right to exist was a major concession, this desire for peace represents a sea change in attitude. Therefore, the long-term prospects for the peace process to move forward again are incontestable, albeit at a frustratingly inconsistent gait. The formal parties to the peace process must deal with the same familiar points—the status of the Jerusalem, Palestinian refugee, and West Bank settlements. All are thorny issues that will test even the most patient and skilled negotiator. However, the less visible and often unmentioned hurdle is water; this issue historically is one of the most difficult to solve in the Middle East, and there is little hope that any peace negotiations between Israelis and Palestinians will be any different. This fortifies many analysts' suggestions that water has eclipsed oil, the source of many previous conflicts, in importance. Ironically, the most oil-rich region of the world is the most likely candidate to host a war, not over the once all-important crude, but over water.

This chapter will specifically address the question of where water will either catalyze conflict or at least complicate peace. It details the significant water interests which Israel, Jordan, the Palestinians, and even Syria must address before a permanent, structured peace can be negotiated in the Jordan River basin. It also models how the current peace proposals would affect water rights, and how those new water allocations and parameters could create the potential for future water conflict. The water issue is a complicated tapestry of interconnected interests and needs; while Israel, Jordan, and the Palestinians obviously are all competing for the same water resources, Israel's water concerns with Syria are complicated by the fact that Syria's water shortages are being exacerbated by Turkey's decision to dam the Euphrates River. Egypt could potentially alleviate some of the water shortages in Israel, Jordan, and the Gaza Strip by diverting water from the Nile, but that would significantly increase tensions with Sudan and Ethiopia. Indeed, creating a properly managed water program in the Middle East that is agreeable to all parties involved will be a very complicated and delicate exercise in negotiations and statesmanship.

Water and the Jordan River Basin Conflict

The conflict in the Middle East is complicated in that it includes several states and nonstate entities: Lebanon, Syria, Israel, Jordan, and the West Bank, as the Jordan (or its tributaries) flows through each of these countries/territories. Water shortages are a common theme in the history of this particular region of the world, often creating a source of conflict. Today access to fresh water in the Middle East, and in the Jordan River basin particularly, is a critical point of survival—water is a life-sustaining resource that is being exploited well beyond the rate it can be replenished. Current meteorological, demographic, economic, and political trends all suggest the water situation in the Jordan drainage area will worsen, at least in the near term. For those who were aware of this increasingly tense hydrography, any euphoria born from the reinvigorated peace process under the leadership of Ehud Barak was tempered; now, it is one more factor adding to the drudgery of the conflict under Prime Minister Ariel Sharon. Indeed, water will be one of the most difficult negotiating points in the peace process and very well may prove too difficult to solve completely. Hence, it is critically important to explore the dynamic between water and the Middle East peace process.

To fully address this question, I will provide a brief summary of the recent political, economic, and military events related to the water issue. Next, I will list the numerous variables affecting the water shortage: hydrology, geology, population growth, economics, and a host of other factors that make water an especially contentious issue. I will then present a brief review of the principal elements of the current peace process, illustrating how water and conflict in the area are inextricably linked. Finally, I will forecast potential events that will highlight how water could derail even the best efforts of well-intended peacemakers and become a proximate cause of conflict in the future.

Recent History

To better understand how water can be a source of conflict in the Jordan basin, a summary of recent "water history" is necessary. Such a brief historical review will illustrate how contested the water issue has been in the region.

Development of water sources in the Jordan basin have been contentious since biblical times, but became especially acrimonious since the Balfour Declaration of 1917, the British document in which Jews were promised a "National Home" in Palestine. The water issue was sparked then because the British declared that if the water resources of the area could be fully developed, at least four to five million Jewish immigrants could move to the area without displacing any of its current Arab occupants. Hence, studies on water distribution began to either corroborate or disprove the Balfour numbers.

Several water studies were conducted resulting in multiple comprehensive water plans and recommendations for the entire region. In 1922, the British published the Mavromatis Plan followed by the Henriques Report in 1928; the World Zionist Organization advocated the Palestine Land Development Company Plan, authored by American J. B. Hayes, in 1935; M. G. Ionides, an engineer and advisor to the Transjordanian government, published an irrigation plan in 1939. This was followed by the U.S.-sponsored Lowdermilk Plan of 1944 and the Anglo-American Committee of Inquiry's Survey of Palestine Plan in 1946. This was the last major plan before the state of Israel formally came into existence in 1947, followed immediately by the outbreak of war.[2] While these plans did have scientific methodologies, their results and recommendations, unfortunately, were more affected by political ideology than by the data collected. If the author or sponsor of the study was pro-Arab, the study illustrated why the land could not absorb Jewish immigrants. If the sponsor was Zionist, the study invariably constructed a more optimistic scenario.

After Israel won its war of independence, it unilaterally began developing water resources. This included draining the swamps of Lake Huleh in northern Israel and construction of the National Water Carrier, both of which enabled Israel to begin exploitation of the Jordan River headwaters. Arabs, outraged at the diversion of water out of the Jordan basin and into the Negev Desert, attempted to divert water from the Jordan River to Lebanon, Syria, and Jordan, before it reached Israeli territory (the project failed for a variety of technical and political reasons).[3] Several minor clashes between Jordanian, Lebanese, Syrian, and Israeli forces, specifically over water diversion attempts, occurred. Indeed, destruction of the other camp's water projects was the main goal in many of these small raids: Syrians ambushed soldiers protecting workers digging trenches to drain the Huleh Swamp in April 1951; Israelis unilaterally began exploiting water sources in demilitarized zones, under heavy (and often-used) military escort. Other minor skirmishes with Jordan and Lebanon occurred frequently as well. Several years later, the skirmishes over water escalated to include the use of artillery, mortars, and aircraft. In February 1964, Israel shelled a joint Syrian-Jordanian water works project designed to exploit water from Lake Tiberias. Israel launched armed raids into southern Lebanon and destroyed three water reservoirs in a border town on October 26, 1965.[4]

Partly because of the increase in tensions over water in the early 1950s, the United States sent a special ambassador to the region. Eric Johnston was the head of the International Council of the Authority for Technical Assistance. He was tasked with creating a comprehensive water scheme for the entire region. He did so, with mixed results. All countries involved generally agreed upon what came to be known as the Johnston Plan. The plan was not signed, however, as Jordan feared such an agreement would necessarily entail a peace treaty with Israel. When tensions between Israel and Egypt increased, leading to the 1956 Sinai Campaign, the Johnston plan was set aside. However, it became the de facto agreement that Arab powers in the region generally attempted to follow.[5] In the meantime, from 1948 to 1964, no fewer than thirteen water development schemes in the region were proposed by various organizations. Since none were fully agreed to, Israel finished construction of its National Water Carrier, a large water carrier which takes water out of Lake Tiberias and the headwaters of the Jordan River, and pumps it to the Mediterranean littoral and the northern Negev Desert.

Hence, although the Johnston Plan remained the de facto water plan for the area for a number of years, as of 1984 Israel exploited approximately 55 percent of the available water in the Jordan basin, although the Johnston Plan allocated far less. This left only 45 percent for Jordan, Lebanon, and Syria.[6]

Today, no comprehensive water plan currently exists, although Arab nations generally still refer to the Johnston Plan when discussing water rights. Only two ratified bilateral water arrangements exist, one between Israel and Jordan, and one between Syria and Jordan. I will discuss these later in the chapter. In the absence of an agreed-upon water framework, the current water situation is only marginally more structured today than it was in the previous decades.

Water and Conflict—The Variables

The water struggle in the Middle East arises from a confluence of variables, including water scarcity, rapid population growth, economic expansion, technological advances, poor water management, ill-defined water laws, and previously existing tensions. None of these variables exists in a vacuum—each affects the others in some way, adding to the complexity of any comprehensive water solution in the Middle East.

Water Scarcity—An Examination of the Water Sources in the Region

It is obvious that the very root of the water problem in the Middle East is a shortage of water. The Middle East is one of the most arid areas in the world. There are three natural sources of water for the subject countries to exploit: river waters, groundwaters, and rainfall. None of these sources is abundant in the Jordan River basin, and what few sources do exist are currently all either fully (or over) exploited, or their use is prevented due to conflict. Other sources of water, which include recycled and desalinated water and the import of water from outside the Jordan basin, are too expensive, too technically challenging, or politically unfeasible, and therefore will not notably affect the peace process, at least in the near term. Desalinization, however, probably holds the greatest promise for alleviating water shortages in the future. An examination of each of the three primary water sources now follows.

River Waters of the Region

There are two major river systems involved in the Arab-Israeli water conflict: the Jordan watershed, comprising the Jordan and Yarmuk Rivers, and the Litani watershed. (Although the Litani is not part of the Jordan basin, it is included in this discussion due to its close proximity to the Jordan and the political considerations surrounding it.) Both Jews and Arabs have claims to these river systems, many of which are irreconcilable and mutually exclusive. The Jordan watershed is especially problematic, as the majority of its headwaters originate in Lebanon, Syria, and Jordan, but is largely exploited by Israel. The Litani poses its own problems, despite the fact it completely originates in and flows only through

Lebanon. Its close proximity to Israel has created ambition south of the border and paranoia to the north.

The Jordan headwaters are comprised of the Hasbani and the Banias Rivers, which originate in Lebanon and Syria respectively. These two rivers converge north of Lake Hula in the northern Galilee in Israel. The Dan River, which rises in Israel, joins the Jordan at Lake Hula. From that point of confluence, the Jordan River flows south through Israeli territory and through Lake Tiberias (the Sea of Galilee). Just south of Lake Tiberias, the Jordan is fed by the Yarmuk River, which originates in southern Syria, and then forms the international boundary between Jordan and Syria before emptying into the Jordan. The southern half of the Jordan, between Lake Tiberias and the Dead Sea, forms the international boundary between Israel and Jordan. Hence, water for the Jordan originates in all four nations, and partially forms an international boundary for three of them. This hydrology complicates political and economic considerations among these riparians. According to a study conducted by the Tennessee Valley Authority in 1953, under the auspices of the United Nations, the total estimated annual flow of the Jordan amounts to 1,880 million cubic meters of water per year (mcm/y), 77 percent of which originates in the three Arab countries, 23 percent of which originates in Israel.[7]

This mixture of water origins, river flows, and international boundaries creates a witch's brew of problems. Over 90 percent of Syria's water is shared with Turkey, Iraq, Israel, Lebanon, and Jordan. Approximately 36 percent of Jordan's water sources are shared with Syria, the West Bank, and Israel. More than half of Israel's water is shared with Syria, Lebanon, Jordan, and the West Bank. With such a complicated hydrological river system, the situation can be volatile. Indeed, the Jordan basin has been described as "having witnessed more severe international conflict over water than any other river system in the Middle East . . . and . . . remains by far the most likely flashpoint for the future."[8]

The Litani River watershed is, geographically speaking, much simpler. It rises in central Lebanon, flows in a southern route completely through Lebanon, and empties into the Mediterranean Sea several miles north of Tyre. Hence, the boundaries of this watershed are completely contained within the political boundaries of Lebanon. However, this has not prevented its waters from being a source of contention in the region. Part of this problem is that the Litani is not yet fully exploited by Lebanon; the extra water flowing unused into the Mediterranean Sea is desperately needed by Syria, Jordan, and Israel. Many speculate part of the reason Israel invaded Lebanon in the early 1980s was to secure, and then partially divert, the Litani. The estimated annual flow from the Litani totals 410 mcm/y.[9]

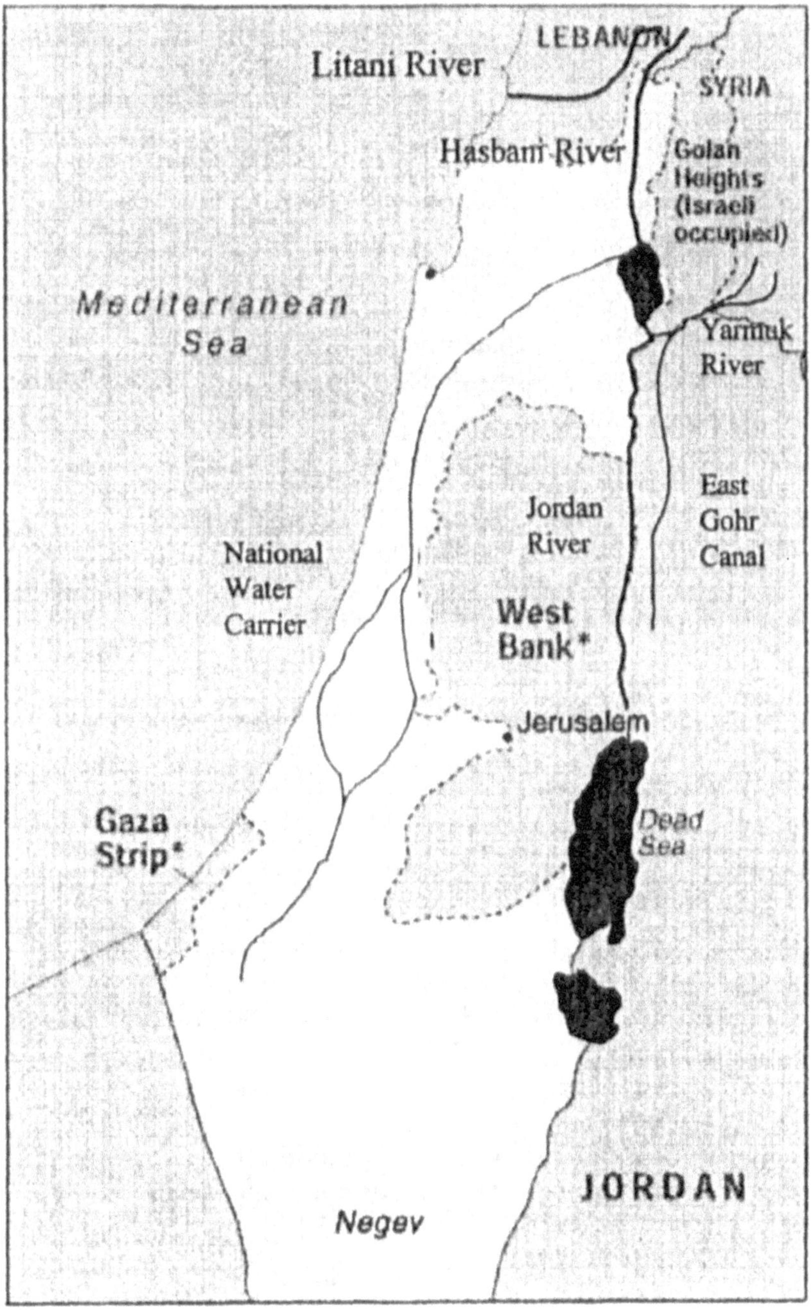

Figure 7.1. Jordan River Basin
Source Modified from A. H. Miller, "The Jordan River—Too Little for Too Many"[10]

Table 7.1. Jordan Flows and Originations

	Source Country	*Gain (mcm/y)*	*Loss*	*Total*
Dan River	Israel	245		
Hasbani River	Lebanon	138		
Banias River	Syria	121		
Jordan (in Hula Valley)	Israel			504
Hula Valley			100	
Local runoff in Hula	Israel/Syria	140		
Flow into Lake Tiberias				544
Runoff/rainfall into Tiberias	Israel/Syria	200		
Evaporation from Tiberias	Israel		270	
Outflow to Lower Jordan				474
Yarmuk River	Syria/Jordan	492		966
Wadis/Springs in Ghor Valley	Jordan/Israel	505		1471

Source: Thomas Naff and R. S. Matson, *Water in the Middle East.*[11]

Groundwaters of the Region

Syria, Lebanon, and Israel all have indigenous underground aquifers within their political boundaries—hence, these particular aquifers are not directly contributing to the Arab-Israeli struggle for water at the sovereign state level. However, the large mountain aquifer located in the West Bank symbolizes the struggle between Palestinian and Jew for access to water.

Groundwater makes an enormous contribution to Israel's total water supply—indeed, Israel uses approximately 850 mcm of groundwater each year, accounting for just under 45 percent of its total water use.[12] However, about 400 mcm/y of this water comes from the mountain aquifer, which was captured from Jordan in the 1967 war. This mountain aquifer feeds springs on both the east and west side of the West Bank; in other words, it flows to the Israeli littoral or into the Jordan. The primary way to access this water from the West Bank is to sink wells, as there are few natural springs directly above the aquifer. Tensions between the Palestinians and Jews arise primarily because Israel refuses to allow Palestinians to sink their own wells in the West Bank, instead insisting the Palestinian villages hook into the Israeli National Water Carrier (NWC). However, Jewish settlers in the West Bank can sink their own wells if they are not yet supplied by the NWC. Hence, Jews can exploit this water, but Arabs are not allowed to do so, creating tensions between Israel and the various citizens of the West Bank. This situation will be further discussed later in the chapter.

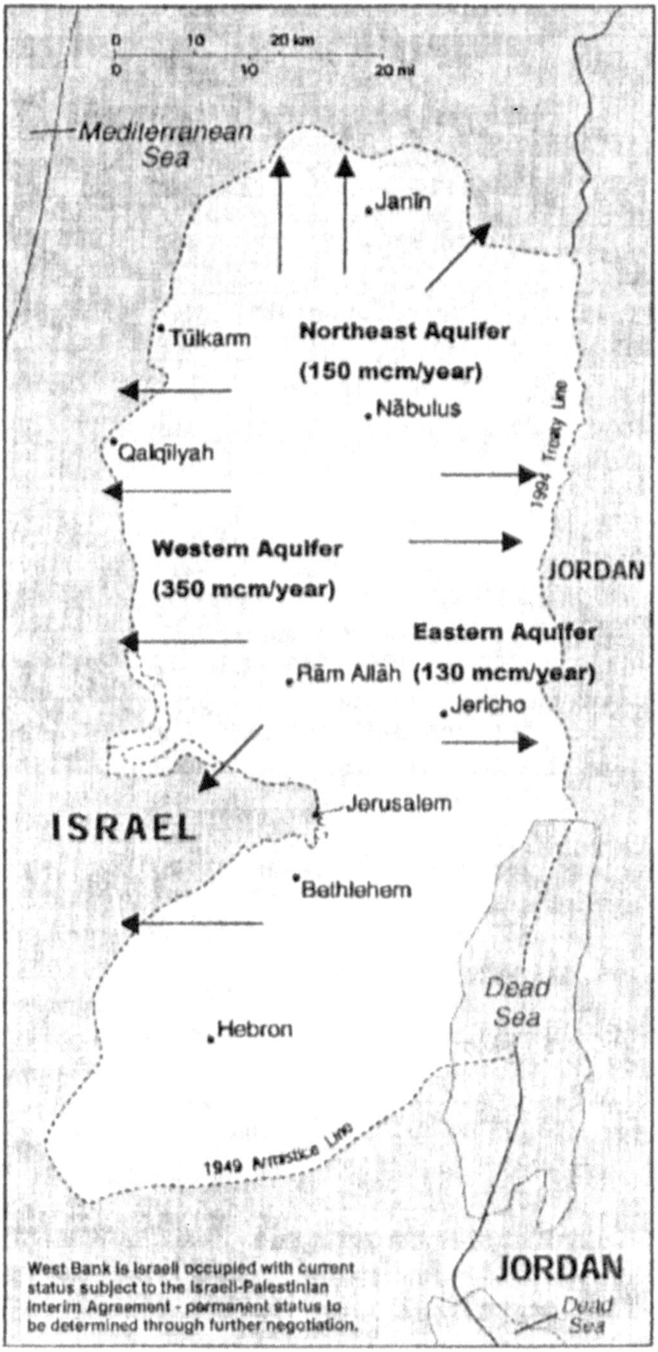

Figure 7.2. Aquifers/Groundwater in the Jordan

Table 7.2. Total Usable Water Availability in the Region, by Source

	River (Surface) Flow (mcm/y)	*Groundwater (mcm/y)*	*Rainfall (mcm/y)*
Israel	600.0	850.0	
Jordan		1.0	
Lebanon	33.0	38.0	120.0
Syria	55.0	70.0	70.0
West Bank	3.0	3.0	

Source: World Resources Institute.[13]

Rainfall in the Region

Rainfall is especially scarce in the Jordan basin, particularly in the exceptionally arid southern Negev Desert in Israel. While northern Israel and the extreme northeast corner of Jordan annually receive approximately 110 cm of rainfall, for both nations rainfall accounts for very little of the region's water, and, therefore, does not supply any meaningful amount of water that can be reliably exploited. Unfortunately, the rain for an entire year in Israel and Jordan usually comes in one or two showers. Syria and Lebanon, on the other hand, receive rainfall in enough amounts to aid in agriculture and some recharging of underground aquifers.

To better put the scarcity issue in perspective, Jordan and Israel are the fifth and seventh driest nations on the planet respectively, both in total water availability and in a per capita measurement. Only Kuwait, Libya, Oman, and Singapore are drier.[14]

Rapid Population Growth—Higher Demand for an Increasingly Scarce Resource

The second greatest factor underlying water tensions in the Middle East is the explosive population growth in the area over the last fifty years. According to the Population Reference Bureau, a Washington, D.C.-based organization that tracks global population trends, the population of the Middle East quintupled from sixty million in 1950 to 286 million in 1996. If the current rate of population growth continues, the population of the region will double yet again in the next thirty years.[15]

Specifically within the Jordan basin, population grew at a faster rate than in the Middle East overall, mostly due to large influxes of immigrants. Syria and Jordan's natural population growth rate is estimated at 3.7 and 3.4 percent, respectively. While Israel's natural population growth is only estimated at 1.5 percent, the continuing influx of Jewish immigrants, especially those from Russia, caused the real annual population growth to be closer to 10 percent since 1987.[16] Additionally, hundreds of thousands of Palestinians moved to Jordan from Kuwait after the Gulf War, further increasing the population of one of the most water-scarce nations in the world.

Table 7.3. Population/Per Capita Water Availability (Population in Thousands, Water in cm/year)

	Israel	*Jordan*	*Lebanon*	*Syria*
1955	1,748/	1,447/	1,613/	3,967/
	1,230	905	3,087	6,501
1990	4,821/	4,259/	2,555/	12,348/
	461	308	1,949	2,089
2025	8,366/	9,369/	5,621/	27,165/
	247–303	104–114	1,021–1,248	713–835
2050	12,549/	11,500/	8,431/	40,747/
	192–300	68–90	768–1,218	454–667

Sources: Falkinmark, World Resources Institute.[19]

As populations increase, the per capita water availability decreases. Population growth in the Middle East is on a trajectory to reduce per capita water availability approximately 50 percent by the year 2025.[17] Indeed, a World Bank Report in 1993 stated, "within one's lifetime, annual average per capita renewable [water] supplies—excluding so-called 'fossil' aquifers in the Arab region—will have fallen by about 80 percent."[18] The outcomes of this increasing gap between supply and demand leads to land and water degradation, disruptions in the hydrological cycle, an increase in waterborne illnesses, and other maladies—all of which will increase insecurity and the likelihood of conflict.

Swedish hydrologist and Population Reference Bureau scientist Malin Falkinmark developed an index to measure the adequacy of water supplies. Her guidelines, which can be compared to the chart above, are: Absolute Scarcity: below 500 cm/y; Scarcity: 500–1,000 cm/y; Stress: 1,000–1,700 cm/y. As the table above suggests, Israel and Jordan were in the absolute scarcity category as early as 1990 in per capita water availability, with projections suggesting that even what meager water there was could be almost halved in twenty-five years. To put this in better perspective, as of 1995, the annual per capita water availability in the United States was 9,413 cubic meters—thirty times that of Jordan and twenty-five times that of Israel.[20]

Rapid Economic Development—Increasing Pressures of Water Usage

Despite recurring war and perpetual tension, Middle East economies have enjoyed significant development in the last twenty-five years. While the benefits of such development are unquestionable, the new pressures placed on scarce water resources are significant.

It is important to note the majority of this growth occurred in the industrial sector of these economies; such industrial growth creates increasing competition for water, which is already being used for domestic or agricultural purposes. Although increasing populations require more food, little growth in the agricultural sectors of these economies (with the exception of Syria) occurred, especially in the last decade. In fact, agriculture constituted 18 percent of Jordan's gross national product (GNP) in 1965, but only 6 percent by 1993.[21] This is mostly because all the water sources available in the region were already fully exploited, preventing any further irrigation of arid lands. Moreover, all lands that do not need irrigation are already fully planted. Hence, growth in these economies is naturally in the nonagricultural sectors: in Israel, high-tech industries are springing up in the Haifa region. Indeed, industrial use of water is projected to continually increase in the Middle East; while it constituted 8 percent of total water use from 1986 to 1993, it is expected to be well over 60 percent by 2003 and 37 percent by 2025. Indeed, in 2000, 75 percent of the country lived in an urban environment—up from 35 percent in 1950.[22]

This increasing industrial use of water creates problems. While industrial uses of water contribute much more to a nation's gross domestic product (GDP) than do agricultural uses of water, industrial runoff is much more difficult to treat and often ends up polluting other sources of water, such as aquifers and streams. This is especially true of areas where water infrastructure is not well developed, as in the Middle East. Hence, pollution of water sources will be increasingly problematic.

Increases in economic growth also typically lead to greater urbanization; over half the Middle East now lives in urban areas where people consume ten to twelve times as much water per capita as village dwellers.[23] For example, in 1960, approximately 35 percent of Jordan's population lived in or around the capital of Amman; in 1991, that figure reached 55 percent, and is projected to be nearly 60 percent by 2000.[24] In 1996, the entire Jordan basin's population was 63.4 percent urban, with that number likely to increase significantly in the coming decade. Thus, economic development created and will continue to drive greater per capita demands for water. Hence, these last two variables (increasing population and a growing economy) create a synergistic spike in water demand.

Table 7.4. Water Use as a Percentage of Total Use

	Domestic	*Industrial*	*Agricultural*
Israel	29	7	64
Jordan	29	6	65
Lebanon	11	4	85
Syria	7	10	83
West Bank	22	0	78
Gaza Strip	15	0	85

Sources: World Resources Institute, 1992; Central Bureau of Statistics, 1998.[25]

Technology Enables Greater Water Exploitation

Modern technology gives humans the ability to alter their natural environment, which is now happening at an unprecedented level and in unprecedented ways. Specifically with water exploitation, technology allows the construction of irrigation canals, dams, and deeper wells. As with economic development, this has both good and bad outcomes. Certainly, the increasing capability to develop previously unusable water with dams and aquifer construction allowed the population of the Middle East to expand, increased regional crop yields, improved sanitation, and enhanced other quality of life factors.

However, this technology is a two-edged sword. Dams create large reservoirs, creating higher water loss due to increased evaporation. Large irrigation projects waste water through inefficient watering. Deeper wells extract larger amounts of groundwater, lowering water tables. Lowered water tables allow saltwater seepage, in turn causing increased salinization of the water table, permanently rendering it unusable.

All these scenarios exist in the Jordan basin. Indeed, water exploitation in the Gaza Strip has caused the salinity of the water table there to reach almost unusable levels. Each year 140 mcm are pumped out, while the natural annual return is approximately 60 mcm.[26] Water projects, to include Israel's National Water Carrier and Jordan's East Ghor Canal, are typical of large water projects which increase evaporation rates. Water usage has so altered the Jordan and Yarmuk Rivers that the Dead Sea (the final basin where the Jordan flows) has dropped by approximately fifty feet in the last eighty years, and its total surface area has shrunk by nearly 40 percent.[27] The Jordan is little more than a rivulet by the time it drains into this ever-shrinking body of water. Hence, technology has enabled the population of the region to exploit water beyond the point of sustainability.

Unequal Distribution of Water

Water rights and allocation is, and always has been, a great source of contention. From the very beginning of Israeli statehood, securing as much water as possible for one's own state or territory was synonymous with survival and denying the enemy water was a way of undermining his strength. Under present water distribution schemes, a very visible disparity in water consumption exists that favors Israel. This is a source of tension both at the international level and between Israel and the Palestinians in the West Bank.

The construction of the National Water Carrier allows Israel to pump approximately 500 million cubic meters per year of high-quality sweet water out of the headwaters of the Jordan and Lake Tiberias.[28] Indeed, this represents the lion's share of the irrigation-quality water in the Jordan. This water, which represents roughly one quarter of Israeli total consumption each year, is completely removed from the Jordan watershed, and pumped to consumers in the Mediterranean littoral and the northern Negev.

This water diversion represents a double blow against Arab water interests in the area. First, by removing water from the headwaters of the Jordan, the flow of high-quality water into Lake Tiberias is lowered. This reduces the quality of water in Lake Tiberias, from which both Jordan and Israel pump additional water. There are salt-water belts under and around Lake Tiberias, which, when the level of the lake got below a critical level in the 1980s, began discharging into the lake itself, further reducing the quality of the lake's water. Israel responded to this by tapping the salt springs and then pumping the saline drainage water downstream to dump it back into the Jordan, where it enters Jordanian territory. Again, since the flow out of Lake Tiberias is reduced, already increasing the salinity of the Jordan below it, the addition of this saline drainage into the Jordan significantly reduces the quality of water Jordan can exploit out of that section of the river. Thus the injury is twofold: Jordan receives much less water from the Jordan River than does Israel, and the water it does receive is so high in salt content it is only marginally useful, even for irrigation.[29]

Ironically, Israel makes water claims on the Yarmuk (Jordan's primary river water source) as well, since the Yarmuk is the primary tributary to the Jordan. This further reduces Jordan's water claim. Indeed, with the severe drought in the region over the past two years, Israel is attempting to reduce the amount of water Jordan can claim from the Yarmuk from 55 mcm/y to 27 mcm/y.[30]

Within Israel and the occupied territories, the uneven distribution of water is obvious as well. In the 1980s, water consumption among Israeli settlers in the West Bank was seven times that of the Arab inhabitants.[31] Recently, Mustafa Natsheh, the mayor of Hebron, indicated the disparity was eight times as much.[32] Western sources suggest a smaller, but still enormous, fourfold difference: Israeli settlers consume between 90–120 cubic meters per year, while Palestinians consume 25–35 cubic meters per year.[33] Furthermore, Israel forbids Arabs living in the West Bank from drilling their own wells or developing any other water resources in an attempt to rectify the consumption imbalance.[34] Indeed, while this ban is in place, Israeli settlers, using the latest technology to drill deep into the aquifer, are exploiting the West Bank aquifer at a rapid rate. Should a political settlement giving the Palestinians control of the West Bank ever be agreed upon, the Mountain Aquifer in the West Bank will be already overexploited, which will make it difficult for the Palestinian Authority (which currently does not have access to the same drilling technology Israel has) to supply adequate water for the remaining residents of the Territories.

Additionally, Palestinians are charged up to twice as much for water as are Israeli settlers—when water is made available to the Palestinians. Some camps in the West Bank have not had water for months. One 7,000-member refugee camp depends on a water supply provided by the United Nations, which has run for just two hours a day since April 2000. Hebron has to truck in water for its 100,000 residents.[35] This creates bitterness and tension between Jews and Arabs, especially when water supply is tight, as it has been in the last two years.

Poor and Difficult Water Management

Water management in the Jordan basin is problematic for two main reasons. The first is the difficulty in exploiting the only outside water that enters the region naturally—rain. Although rains in the area are seasonal, they are very unpredictable. The majority of the rain typically comes in the winter months, leaving the hot summer months with little to no precipitation. Moreover, the rains are often irregular and localized, making their capture and exploitation nearly impossible. Thus, while Jordan and the West Bank receive on average 20 mcm/y of rainfall, only 3 mcm/y of that amount can be captured and utilized, because it comes in only one or two rainfalls. Furthermore, due to the extreme arid nature of the region, high evaporation and transpiration rates diminish the amount of rain that can be used.

On a strategic level, none of the countries under discussion use their water efficiently. Israel, Syria, and Jordan all insist, as a matter of national policy, that they need to be as self-sufficient in food as possible. Thus, to paraphrase the first Israeli prime minister, Ben Gurion, they try to make the desert bloom, despite the extraordinarily heavy water burden this creates.

Israel uses 1,180 mcm (62 percent of total supply) and Jordan uses 67 mcm (74 percent of total supply) for agriculture. Because the agricultural lobby is so strong in both governments, each heavily subsidize water. While both nations have done much to increase the efficiency of their irrigation methods (indeed, Israel was a pioneer in the development of drip irrigation technology, and boasts only a 12 percent loss during irrigation[36]), the contribution of each nation's agricultural sector to GNP has steadily declined over the years. In 1993, only 2.6 percent of Israel's and only 6 percent of Jordan's GNP was generated by agriculture.[37] These figures suggest that while both Israel and Jordan are improving their water efficiently within their agricultural economic sectors, a more appropriate overall water strategy would be to import more food and transfer the newly available water to efficient industrial and domestic uses. This would not only boost GNP (a ton of water in industrial uses typically generates $14,000, while a ton of water in agriculture typically generates $1,000[38]), but would also leave additional water for inevitable future population growth in the region. Interestingly, due to the perceived vulnerabilities both nations argue would exist due to the resulting food dependence, such a change in strategic water management is highly unlikely. The political clout of the agricultural sector in each nation also prevents any significant reduction in subsidies.

International Law—Muddling the Issue

Existing international law regarding water sources shared among multiple sovereign states is vague and does not apply well to developing areas. The majority of international water law was developed in Europe and is concerned mostly with navigation rights and water purity, rather than water availability.[39]

Hence, international law provides little concrete guidance to the Jordan basin riparians.

Two conflicting legal principles exist concerning international freshwater laws: sovereignty and integrity. Absolute sovereignty is advantageous to the upstream riparians as it regards water as an integral part of a state's national territory. This doctrine states, "Territorial sovereignty is the sovereignty applied to a specific country or basin, the right to make decisions with regard to his country or basin without having to consult other countries and their citizens."[40] Downstream riparians who may object to the way the upstream riparians are developing common water sources are considered to be meddling in the internal affairs of the sovereign nation where the water originates. This is the principle Turkey used to create the massive reservoir behind the Attaturk Dam in 1989. Despite earlier guarantees of a constant flow of water, Turkish president Turgut Ozal rattled Syria and Iraq by deciding to hold back the flow of the Euphrates for a month. Indeed, in November 1998 Syria threatened to bomb several of the smaller dams that make up Turkey's Grand Anatolia Project (GAP), a massive hydrology project that reduced the Euphrates' flow by approximately 60 percent. The Turks responded that the water originating within their national borders was as much theirs as oil within the boundaries of Arabic nations belongs to the Arabs. If Arabs, despite having used the waters of the Euphrates for centuries, wanted to make a water claim against Turkey for the GAP, then Turkey would have the right to make oil claims against downstream Arab riparians.

The second legal principle is that of absolute integrity, which focuses on the integrity of the body of water rather than the sovereignty of the state where the body of water originates. This legal doctrine states that

> a river which flows through the territory of several states or nations is their common property. . . . Neither nation can do any act which will deprive the other of the benefits of those rights and advantages. The inherent right of a nation to protect itself and its territory would justify the one lower down the stream in preventing by force the one further up from turning the river out of its course, or in consuming so much water for purposes of its own as to deprive the former of its benefits.[41]

This principle obviously favors the downstream riparians.

When these two principles come into conflict, they do not permit a reconciliation of interests—indeed, they are mutually exclusive. Historically, the outcome of such a conflict of legal principles is dependent on a single factor—which state advocating a particular doctrine is stronger.

The Jordan basin is very complex legally speaking and would be so even if well-developed water laws existed. As mentioned earlier, the lion's share of the Jordan River's water rises outside of Israel, in Lebanon or Syria. Israel, through military action, subversion, and negotiation, has been very successful in preventing any attempted diversion of these waters before they enter Israeli territory. This allows Israel's National Water Carrier to extract almost 70 percent of the Jordan River after it enters Israel, but before it reaches a point where the Palestinians in the West Bank can exploit it. Indeed, in 1967, Israel shelled a joint Syrian/Jordanian dam on the Yarmuk River, as it viewed the dam as a threat to its water supply. Both situations illustrate Israeli reliance on the absolute integrity principle—despite the water originating outside its political borders, Israel makes a claim to the water. However, when addressing Palestinian complaints of not leaving enough water in the West Bank aquifer or Jordanian complaints of needing more water flowing out of Lake Tiberias, Israel follows the principle of absolute sovereignty. Hence, Israel is implementing both (albeit contradictory) legal principles to maintain its share of water.

The matter is even more complicated than this. Of the approximately 1,950 mcm/y of water Israel uses,[42] roughly 950 mcm/y, or 48 percent, comes from territory Israel captured during the 1967 war: 400 mcm/y from groundwater in the West Bank and 450 mcm/y from the upper Jordan. Depending on the legal status of the Occupied Territories and the Golan Heights, the legal status of water ownership becomes even murkier. As later discussion will point out, the current possibilities of returning the Golan to Syria and granting greater autonomy to the Palestinians in the West Bank will greatly increase the contention of which population can legally exploit those 950 mcm/y of water.

As noted, only two international water agreements exist in the region: the Jordanian-Syrian Agreement on the Utilization of the Waters of the Yarmuk River (signed in September 1987) and the water provisions in the Peace Treaty between Israel and Jordan (signed in October 1994). Lebanon has no water treaties with any of its neighbors, and the two most bitter enemies in the region, Syria and Israel, have no bilateral water agreements. Because of the complicated nature of the water system in the Jordan basin, creating more than bilateral agreements is difficult. Also due to the complicated nature of the water system, a comprehensive water agreement among all four nations is necessary to achieve a lasting solution to the water crisis in the area.

Previously Existing Tensions

Water use is simply one of many tensions between the peoples of the Jordan River basin. It has been the primary source of conflict in the past, as both Arab and Jew have attacked the other's water infrastructures based on perceived threats. It has also been the proximate source of conflict, as many Arab analysts claim one of the overriding reasons for Israel to go to war in 1967 was to capture additional water sources.[43] But even if there were adequate water sources in the Middle East, there would still be plenty of reasons for conflict to exist. As Israeli hydrology professor Uri Shamir stated for *National Geographic*, "If there is political will for peace, water will not be a hindrance. If you want reasons to fight, water will give you ample opportunities."[44]

The loss of water sources to Israel seems to be a national shame to Palestinians, Jordanians, and Syrians. Priit Vesilind, a reporter for *National Geographic*, conducted a series of interviews in 1993 throughout the Middle East to gauge sentiment on the water issue. A Syrian border official in the Golan Heights stated in 1993, as he pointed to an Israeli flag flying across the demilitarized zone, "See that flag? Those blue stripes represent the Nile and Euphrates. The Israelis think this is where their land should extend, all the way from Egypt to Turkey. And they are working to get this area." Abdullah Toukan, the science advisor to the late King Hussein of Jordan suggested, "In this arid region water is life. Money may bring desalination plants, but the real solution remains the restoration of Jordan's rightful share of water." Palestinian elder Shaher Khufash, living in the village of Marda in the West Bank, once stated, "Israel has stolen our water, and we are thirsty." Despite the fact Israel built National Water Carrier infrastructure to the village, the village elders refused to hook up to Mekorot, Israel's national water supplier. "Doing so means accepting them and their confiscations. Also, they say we have to pay for the water then. The people refuse."[45] It seems from these limited statements that at least some believe the solution to current water shortages isn't to make sure everyone has enough water; rather, it's to make sure your enemy doesn't have any water.

Until recently, Israel maintained a security zone in southern Lebanon. Katyusha rockets, fired by Hizbollah extremists, slammed into northern Israel regularly until the Israeli pullout in June. Israeli tanks still practice maneuvers in the Golan Heights, anticipating a Syrian attack. Syria will not even discuss peace with Israel unless Israel comes to the negotiating table willing to return the Golan Heights. A peace treaty between Israel and Jordan was signed only five years ago; water rights were the last issue to be agreed upon, and are among the main issues threatening to undermine the otherwise workable pact. And although the Camp David Accords sealed a peace treaty between Israel and Egypt decades ago, the peace is an uneasy one. In this environment of mutual distrust, terrorist organizations, occupation zones, demilitarized zones, occupied territories, and a history of a major war every decade since 1947, there is little wonder progress on such things as cooperation in water use hasn't received more attention.

The Current Situation and Water

While the Knesset frequently debates the necessity of a regional water solution, Israeli policy makers have generally chosen to "go it alone" in terms of water management. Without a formal regional peace agreement, a formal regional water solution would be exceptionally difficult, despite the numerous and varied attempts. With Israeli economic and military might unchallenged in the area, the unilateral water policy Israel pursued could not be challenged by the Arabs. However, when a permanent peace is established, that must change—Israel's "imposed" water solution to the region could not survive.

While the details of any peace talks that may be occurring between Israel and the Palestinians (and last year, Syria as well) are unavailable, the most important issues are certainly the same—the status of Jerusalem, status of Palestinian refugees, and Israeli settlements. Prime Minister Sharon of Israel, unlike his predecessor, is not willing to negotiate land for peace; he is staunchly opposed to halting future settlements, will not consider the return of the Golan Heights to Syria, or give greater autonomy (and possibly allow the Palestinians to declare statehood) in the West Bank. However, future Israeli leadership may return to these ideas; indeed, the Israeli population is already showing signs of fatigue with the hard-line approach chosen by Prime Minister Sharon. These actions, should they eventually be adopted, will have significant ramifications for the balance of water in the region. A discussion of each follows.

Return of the Golan Heights

By returning the Golan Heights to Syria, Israel will relinquish its nearly exclusive control over the headwaters of the Jordan River. This is a major sacrifice, one that would be difficult (although not impossible) for Israel to make. According to Thomas Stauffer, a former professor at both Harvard and Georgetown Universities, Israel's control of the Golan is significant to its water supply for three reasons.[46]

First, control of the Golan prevents Syria, Lebanon, and Jordan from developing the headwaters of the Dan, Banias, and Hasbani Rivers. In the 1960s, these three nations developed their own unified water plan that essentially would divert water from these rivers before it ever entered Israeli territory. As mentioned earlier, Israel bombed these areas to prevent such a plan from ever reaching fruition. Relinquishing the Golan Heights now would at least strategically allow these three nations to pursue development of those headwaters. Should these Arab states choose to do so, a significant reduction in water to the upper Jordan, and consequently to Israel, would result.

Second, Israel has expressed an interest in including the Litani River in southern Lebanon in any regional water plan. Indeed, some articles suggest Israel's invasion of Lebanon in the 1980s was to secure the Litani for itself; the fifteen-year maintenance of the security zone only fed those fears. Control of the Golan protects Israel's flank should it ever decide to divert Litani waters to the

Hasbani River, thus increasing the flow of water into the headwaters of the Jordan. While this is now highly unlikely due to Israel abandoning the Lebanese security zone, it is not completely out of the realm of possibilities. Should the Hizbollah or even Syria begin launching raids against northern Israel from Lebanon again, a more hawkish Israeli government might reinvade, and seriously consider such a water diversion plan. By giving up the Golan now, this will preempt any such action regarding the Litani by the Israelis.

A third factor in giving up the Golan is the protection of the inlets of the National Water Carrier and the pumping stations from possible bombardment of Syrian artillery. Again, this scenario is unlikely (Bashar al-Asad doesn't appear eager to militarily engage Israel), but there is no guarantee given souring relations between Israel and Syria regarding Prime Minister Sharon's current policies. Should Syria place military forces in the Golan, the entire Hula Valley and Lake Tiberias would be in range of Syrian artillery.

In short, giving up the Golan will make Israel vulnerable to Arab designs on Jordan River water, complicate any future Israeli water designs on the Litani, and expose Israel's northern tip once again to Syrian military forces. No doubt, this will be a major point in the negotiations between Israel and Syria. Indeed, the most contentious (and last) issue to be agreed upon in the peace talks between Israel and Jordan in 1994 was the creation of a Joint Water Committee to develop additional water resources on the Yarmuk River. The Yarmuk is much less significant to Israel than are the feeder rivers to the headwaters of the Jordan, which will likely make this aspect of any peace agreement especially difficult.

Increased Palestinian Autonomy in the West Bank

Israeli hegemony in the West Bank is crucial to Israel's water supply. As previously mentioned, approximately 400 mcm/y of water from the West Bank aquifers feed into Israel's National Water Carrier. Additionally, Israelis have, and must continue to, preempt local use of the mountain aquifer so that the water continues to flow to Israeli surface water sources along the coast, rather than be siphoned off by residents of the West Bank drilling wells. Should Palestinian wells be sunk in the West Bank, the aquifer could quickly be exploited to the point Israeli springs in the Mediterranean littoral would go dry, or at least experience significantly decreased flows.

Additionally, there is the concern of water pollution. Palestinian water projects are primitive compared to Israeli standards. What little agriculture they do have uses inefficient irrigation techniques, and the black market wells that do exist add to the pollution of the aquifer. Indeed, some sewers in the West Bank simply empty straight into the aquifer. Should the Palestinian Authority (in whatever form, be it an autonomous controlling authority or the national government of the state of Palestine) take more control of the water sources in the West Bank, the Israeli springs and wells fed by the West Bank aquifers will almost certainly become more polluted.

Immigration would also pose a significant impact on the water situation if Palestinians assume control in the West Bank and their people are allowed to return home. Approximately two million Palestinians live around the world—a de facto Palestinian diaspora. Should a Palestinian State become reality, these two million people would have a politically recognized homeland based on their ethnicity. Just as Israel encouraged millions of Jews to immigrate after declaration of statehood (in not too dissimilar circumstances) in the last five decades, hundreds of thousands of Palestinians, both in the refugee camps in Jordan and in the Palestinian diaspora, may decide to move to the West Bank. Such an influx of new residents would create an incredible strain on an already marginal water supply. Assuming Kliot's assessment (previously quoted) that each Palestinian consumes approximately twenty-five to thirty-five cubic meters of water per year, one million Palestinian immigrants to the West Bank would create an immediate additional need of twenty-five to thirty-five million cubic meters of water per year—a demand for which there is currently no supply.

Israeli Withdrawal from the Security Zone

Israel's withdrawal from the Security Zone in Lebanon now makes it extraordinarily difficult for Israel to make any claims on the Litani River in the future. Indeed, increasing populations in both Israel and the West Bank may make this an eventual necessity, as the Litani is the only water source in the region that is not close to being completely exploited. However, without Israeli hegemony (if it ever really existed) in southern Lebanon, such exploitation of the Litani would certainly need to come about through negotiations with the Lebanese and Syrian governments, rather than force.

Peace, the 1967 War, and Conquered Water

The 1967 war vastly increased Israel's supply of water. The possible loss of this water due to territorial concessions would create a crisis in Israel—a crisis that would be exceptionally difficult to solve. There is no new water in the Jordan basin area that is waiting to be developed. As early as 1981, Israel was exploiting 99 percent of its available water. Drought, like in the early 1980s and since 1997, has forced Israel to critically examine the value of the water it does use. War in 1967 made up the water deficit Israel needed to continue to expand economically and agriculturally, as well as to continue absorbing Jewish immigrants. The peace process strongly suggests Israel will now have to share that water it once conquered. Where Israel will acquire new water to replace the water it gives up in the peace process is unknown.

Table 7.5. Significance of Water Captured by Israel in 1967 War

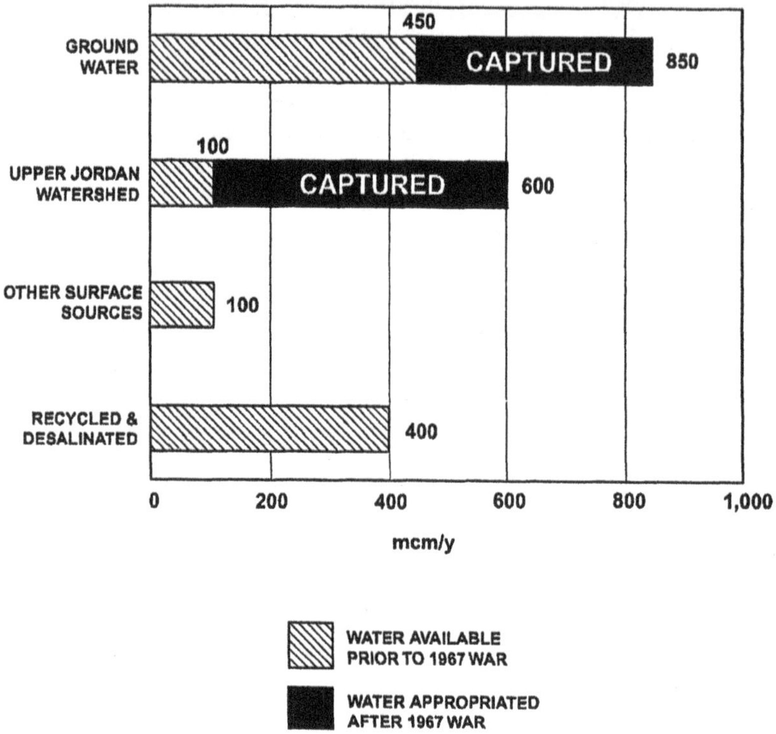

Source: Stauffer, Center for Policy Analysis on Palestine, 1996.[47]

This factor alone will be one of the most arduous issues on the negotiating table. Any peace with the Palestinians must address increased water rights for West Bank Palestinians. A sovereign West Bank would complicate further exploitation of the Jordan. The waters of the Golan Heights, while not important to Syria prior to the 1967 war, are now increasingly significant due to Turkish exploitation of the Euphrates River. Hence, Syria will be very reluctant to share any water originating from the Golan with Israel should a peace treaty be signed. Negotiating these water issues will be extraordinarily problematic.

In sum, Israel's imposed water solution to the Jordan basin cannot survive the peace process unless Israel is ready to abandon its agricultural sector. Only by thinking in terms of a regional water solution will Israel be able to meet current water demands and effect peace in the region.

Table 7.6 dramatically depicts the water challenge Israel would face by returning territories occupied in the 1967 war. It is important to note that this is available water, currently used and exploited. The chart shows how much total water Israel would lose control of, should the Golan and the West Bank be returned.

Table 7.6. Jordan Flows/Originations, Post Peace Process

	Source Country Pre/Post Peace	*Gain (mcm/y)*	*Loss*	*Total*	*Israeli Loss*
Dan River	Israel to Syria	245			245
Hasbani River	Lebanon	138			138*
Banias River	Syria	121			121*
Jordan River (in Hula Valley)	Israel			504	
Evaporation in Hula Valley	Israel		100		
Local runoff	Israel/Syria to primarily Syria	140			Approx.
Flow into Lake Tiberias	Israel/Syria			544	
Runoff/rainfall into Tiberias	Israel/Syria	200			
Evaporation from Tiberias	Israel		270		
Outflow to Lower Jordan	West Bank/Jordan			474	
Yarmuk	Israel/Syria/Jordan	492		966	
Wadis/Springs in Ghor Valley	Jordan/Israel to Jordan only	505		1,471	505**
TOTAL ISRAELI LOSS OF WATER CONTROL				**1,109**	**1,109**

Source: Naff and Matson, 1984. Modified by author, 2000.[48]

Notes:

* Although the Hasbani and Banias originate in Lebanon and Syria respectively, Israel's control of the Golan allows Israel to exploit them. Both these rivers run almost exclusively through the Golan before they flow into the Jordan. Hence, the loss of the Golan would prevent Israeli exploitation and (geographically, at least) enable Lebanon and Syria to begin exploitation of these headwaters prior to their flow into Israel.

** From the loss of the West Bank.

The total loss of water control is higher than Stauffer's water loss figure, as every single cubic meter of water in the Jordan is obviously not used.

Future Water Conflicts

Conflict over water in the Jordan basin remains possible, despite the current prospects for regional stability. As demand shows no sign of slowing and new supply is not available, in the near term the water situation can only become more critical. Indeed, the return of the Golan and loss of the West Bank aquifer would have profound effects on the geohydropolitical structure of the Jordan basin.

Theoretical Framework to Analyze Potential for Conflict

The Middle East Research Institute has directed an ongoing international study of water in the Middle East since 1984. The thirteen-volume set includes a theoretical framework from which one can roughly predict the potential for conflict. To simplify the data, the authors constructed a matrix that predicted the relative power and conflict potential in the Jordan basin. This matrix was constructed on three factors: perceived water need, power, and riparian position. These factors were then added to get a total that tentatively reflects the relative strength of the individual riparian (see table 7.7).

According to the model, the greatest potential for conflict exists when a lower riparian is a more powerful actor than the upper water controlling riparian and perceives its water needs to be deliberately frustrated. Additionally, the model suggests that a significant geohydropolitical inhibitor to conflict is when the uppermost riparian is the most powerful actor in an international river basin.

Should Israel, fully adopting a "land for peace" paradigm, actually return the Golan Heights to Syria, this model suggests an ironic outcome: the peace process will create the first condition and remove the second condition, setting up the most likely scenario for conflict in an international river basin. Israel, having withdrawn from the security zone in Lebanon, and if it gives up much of the headwaters to the Jordan by returning the Golan to Syria, would essentially become a lower riparian state on the Jordan River. The power differential between Israel and the new upper riparians is enormous, thus satisfying the requirements for the most unstable river basin arrangement: that of a lower riparian being more powerful than the upper riparians who control the water source. Hence, the peace process, at least according to this theoretical construct, will structure the Jordan River basin in such a way as to maximize the likelihood of conflict in the future.

If the Palestinians increase their control over water resources in the West Bank, the zero-sum nature of the situation will doubtlessly make many Israelis think the Palestinians are frustrating Israeli attempts to meet its water needs. Again, given the power differential suggested by Frey's matrix, this only increases the likelihood of (continued) conflict between Israelis and Palestinians.

Table 7.7. Relative Power as a Measure of Conflict Potential

	Perceived Need	*Military Power*	*Riparian Position*	*Total*
Israel	5	9	5	19
Jordan	5	2	2	9
Syria	3	3	2	8
Lebanon	1	0.5	2	3.5

Source: Frey, 1992.[49]

This discussion highlights well why water is the thorniest of issues in the peace process. Without Israeli concessions on water, there can be no real progress toward peace; the most obvious concessions Israel could make are the very things that, at least according to this model, will be the source of conflict afterwards.

Specific Indicators of Potential Conflict over Water

While a theoretical framework is certainly helpful in assessing the potential for conflict in a given region, it does little to actually help predict when that conflict will occur. Given the hydropolitical situation in the Jordan basin, and knowing what specific factors contribute to the current water crisis, it is possible to forecast some specific indicators of impending water conflict. At least three specific developments could occur which would, once again, create conditions ripe for interstate conflict over water.

High Level of Palestinian Autonomy/Sovereign State in West Bank

Any increase in Palestinian authority in the West Bank will proportionally increase tensions with Israeli settlers. The greater the authority and autonomy, the more likely clashes between Palestinians and Jews will be. While the rise in Palestinian authority will be the prime reason tensions and the possibility of conflict will increase, water would be a very convenient and likely proximate cause of any outbreak of violence. Indeed, as many Israeli settlers in some way depend on local water infrastructure instead of exclusively on the NWC, any attempt by Palestinian authorities to curtail directly the West Bank Israelis' consumption of water or increase their own consumption to the detriment of the settlers is likely to be a catalyst for conflict.

This is a very likely outcome of the peace process. Skirmishes between Palestinians and Israeli settlers in the West Bank are likely to be local and contained should the Palestinians simply increase their authority over water. Again, water would not be the primary source of conflict—rather, any perceived control or superiority by Palestinians over Israelis would be the primary source. Water is simply a likely catalyst that would ignite nationalistic passions.

However, full-blown Palestinian statehood may cause much more widespread violence. Declaration of a Palestinian state will, without a doubt, cause a great deal of consternation among many Israelis, especially those who may have to move out of the West Bank, either by choice or coercion. Any remaining Israeli settlers would be affected by a sovereign Palestinian government, which at the very best would treat the Israelis and Palestinian citizens equally. There is no reason to believe this equality would not include water allocations. Hence, Israeli consumption of water would be cut by at least 75 percent (assuming Western sources are the most accurate concerning the distribution of water in the West Bank), possibly by as much as 88 percent (assuming Arab sources are the most accurate). However, if Israelis are not treated equally with Palestinians, their water supply may very well be cut completely. In either case, conflict throughout the West Bank between Israeli settlers and Palestinians should be assumed.

Massive Immigration from the Palestinian Diaspora to a New Palestinian State

As previously mentioned, the creation of a sovereign Palestinian state may result in massive immigration to the West Bank. If only half of the Palestinian diaspora moved to the West Bank, an immediate additional need of 25–35 mcm of water annually would exist. This would quickly deplete the already damaged West Bank aquifer and the Jordan River, in turn creating a need for new water resources outside the region within a few years. Additionally, the underground flow of the West Bank aquifer to springs in the Israeli littoral would be significantly reduced, further curtailing Israeli water supplies. The remaining flow of the Jordan would also need to be shared between the Palestinians and Jordan. In essence, the artificially introduced and very sudden increase in population would put a tremendous strain on already faltering water resources.

Such a strain on these resources probably would not result in large-scale conflict between Israel and the newly created Palestinian state. However, general tensions in the region would rise. An independent Palestinian State would be dependent on Syria, Jordan, and Israel to ensure adequate flows from the Yarmuk and Jordan Rivers to get their necessary water supplies from the Jordan River. Jordan and Palestine would need to come up with a water sharing agreement for the Jordan River, an agreement that would almost by definition reduce Amman's already meager supply of water. A nascent Palestinian state would have no military power, and only traces of economic influence. Syria and Jordan, not just Israel, may not see it in their best interest to be generous with such a scarce resource. Hence, the tension would not be just between Arab and Jew, but between Arabs as well. The likelihood of armed conflict between these states is low—but that is partially because the Palestinians would not be able to use force to ensure their share of water.

Israeli Government Cuts Water Supply to Palestinians in Gaza and West Bank

Prime Minister Sharon adopted a much more militant course of action when dealing with the Palestinians. While low-level peace talks are continuing, few believe they will produce a breakthrough anytime soon. Indeed, Sharon has shown a penchant to order military activities against the Palestinians which clearly affect noncombatants, to include using bulldozers to raze portions of Palestinian neighborhoods and even raze farmland and orchards. Given these current tactics and Sharon's implication in the massacres at the Palestinian refugee camps of Sabra and Shatila in September 1982, it is not unreasonable to think Sharon may consider using water as a source of leverage in the current conflict. Conflict is almost certain to intensify and become even more bitter in such a case.

Summary

Water has always been a source of contention in the Middle East, and the current peace process will not change that. Indeed, all of the contributing factors that caused the water situation in the region to reach critical proportions are becoming worse, not better. While the Barak government certainly had good intentions in creating a lasting peace in the region, the concessions he was considering making to the Syrians and the Palestinians not only proved to be overwhelmingly unpopular among Israelis, but may have served to ensure conflict over water in the future. While the current conflict is dimming the very notion of peace, let alone obscure hurdles to peace, establishing water protocols will be an exceptionally difficult maze of competing demands from a variety of uncooperative parties. Indeed, the longer a peace agreement takes, the greater the demographic challenges which elevated water to such an ignoble status will become.

All the factors discussed in this chapter highlight the urgent need to develop a regional water strategy that is accepted by all parties in the Middle East. Whether these strategies include the introduction of outside water to the Jordan basin, technological advances (such as extensive desalination and large increases of water reuse from domestic and industrial waste) or a decrease in agricultural use in water, cooperation among all the actors in the Jordan basin is absolutely critical. The peace process, as difficult as it may be, is the first step toward that cooperation.

Policy Implications

The international community is taking human security, and thus environmental security, more seriously each year. Most notably Australia, New Zealand, and Norway are leading the way in arguing for these new notions of security, focusing on the security of the individual that ignores international boundaries, rather than the traditional state-centric focus on security issues. It is no coincidence that these three nations suffer from environmental problems: ozone depletion in Australia and New Zealand, and radioactive fallout from Chernobyl in Norway. It is only a matter of time until more states, to include the United States, are more significantly affected by global environmental matters. While the United States is slowly moving away from a state-centric notion of national security (as evidenced by the 1999 White House National Security Strategy document, which indicated disease and a degraded environment are U.S. security concerns[50]), it needs to do so more quickly and become more fully engaged in human security efforts around the world.

The theoretical framework used in this chapter, developed by the Middle East Research Institute, is a good (although simplistic) first step in developing predictive tools where the environment could contribute to conflict. By measuring riparian position, perceived need, and military strength, a security analyst can quickly approximate how free a government may feel to use force against another state over water issues. A stable situation exists where the most powerful state is also the state that controls the headwaters of an international river basin. The most unstable situation exists where the strongest state controls none of the headwaters, but feels its water needs are being aggravated by a weaker (and thus easily defeated) upstream riparian.

Whether or not this model is completely accurate is secondary to the fact it certainly highlights: water issues in an international river basin are exactly that—international issues. Thus, water programs should not be developed unilaterally. With water demand on the rise and water supply waning, water programs should be developed on a basin-by-basin level at a minimum.

In the Jordan Basin, water is a traditional source of conflict, and promises to be even more contentious in the future. Currently, Israel is the undisputed power in the region, and controls the headwaters of the Jordan. However, if Israel does return the Golan Heights to Syria, coupled with the withdrawal from the security zone in Lebanon, Jerusalem will no longer control the headwaters of the Jordan. To exacerbate the situation, any Palestinian state in the West Bank would challenge Israel's exploitation of the aquifer below. The peace process would thus take the Jordan Basin from the most stable to the least stable situation.

While the water issues described above are certainly not global in nature, they are international. Decisions made by the Turkish government concerning the flow of the Euphrates will affect Egypt's decisions on water uses for the Nile. The Jordan Basin, in between these two basins, is affected by both. As such, a truly comprehensive water plan, including the waters of the Tigris, Euphrates, Litani, Jordan, and Nile Rivers must be negotiated by all fifteen states involved. Until an international agreement and cooperation on these disputed waters is in place, the potential for conflict and violence will remain high. As such, it is imperative international organizations such as the United Nations have a role in such agreements. Since the United States is already playing a key mediation role in the Middle East peace processes, in addition to having Israel, Egypt, and Turkey as strategic allies, Washington should necessarily be involved in any international water agreement in the region.

Such a meta-state approach to water management is appropriate. Like air, water is truly a global, common resource—actions of one state concerning a river (be they benevolent or harmful) will necessarily affect the capacity of neighboring states to act. Currently, the United Nations various aid organizations and environmental interest groups follow water issues. In the United Nations specifically, water issues are followed under at least the Sustainable Development, Human Settlements, Social Development, Humanitarian Affairs, and Environmental Programs. There is no single international body of water governance. Such a body ought to be considered to help develop international water law and mediate international water issues, help spread water-saving ideas, and generally be a forum to address issues such as the ones presented in this chapter, and the dozens of other potential water conflicts around the world.

It is important for any organization involved in promoting peace and stability in this region of the world to understand the growing significance of water. With a fixed supply and a skyrocketing demand, access to this fundamental resource will become increasingly important to governments and populations. Developing expertise in water management is the first step in cultivating the skills to negotiate conflicts where water is an issue, and it will be the cornerstone of any attempt to create a comprehensive water plan. The scientific, economic, geographic, sociologic, and political communities need to focus increasing attention on developing data, models, theories, and histories of water use and needs. The United States and United Nations ought to encourage such efforts.

The United States should also take a leading role in promoting the spread of water-saving technology. For example, because Egypt continues to rely on traditional ditch irrigation, it uses far more water than is necessary to grow its food. Should Egypt adopt Israeli-developed drip irrigation, the agricultural demand for water would decrease by orders of magnitude. This would free up tremendous amounts of Nile water for other uses and for other states, thus reducing tensions. Furthermore, the United States should encourage both U.S. and foreign firms to develop desalinization technology, and promote its use around the world. While Saudi Arabia leads the world in desalinized water use, it is still economically unviable for most nations. Israel is developing new techniques to desalinate water that is less costly, but it is still not cheap enough for widespread use. Finally, the United States should encourage the sharing of energy-saving technology in order to reduce the need to build dams for hydroelectric power. By allowing lesser-developed states to "leapfrog" to more efficient energy infrastructures, the need to alter rivers and water usage for energy-related purposes could be avoided.

Peace depends on satisfied states and populations, generally free of external coercion or threat. Adequate food, shelter, clothing, medical care, and education are key components to keeping populations healthy and satisfied. These are the basic components of human security. At the very core of each of these things is the need for water. The more contentious this basic component for life becomes, the more likely populations will fight over it. And it is becoming very contentious indeed.

Notes

1. Madeleine Albright, remarks at event sponsored by the World Resources Institute and National Defense University in recognition of Earth Day, Fort Lesley J. McNair, April 10, 2000, Washington, D.C., http://secretary.state.gov/www/statements/2000/000410.html (as of August 10, 2000).

2. Elisha Kally, *Water and Peace: Water Resources and the Arab-Israeli Peace Process* (Westport, Conn.: Praeger, 1993), 5–13.

3. Christine Drake, "Water Resource Conflicts in the Middle East," *Journal of Geography* (January/February 1997): 4–11.

4. David Paul, "Water Issues in the Arab-Israeli Conflict," http://www.salam.org/palestine/water.html (as of August 10, 2001).

5. Kally, *Water and Peace*, 16–19.

6. Hassan Bin Talal, Crown Prince of Jordan, address to International Arab/Israeli Water Symposium, Amman, Jordan, February 25–26, 1994.

7. Charles T. Main, "The Unified Development of the Water Resources of the Jordan Valley Region," prepared at the request of the United Nations under the direction of the Tennessee Valley Authority, Boston, 1953.

8. E. Anderson, "Water: The Next Strategic Resource," in *The Politics of Scarcity: Water in the Middle East* (Boulder, Colo.: Westview Press, 1988), 3.

9. Salah Halawani, "Lebanese Development Projects and Israel's Pursuit of the Litani and Hasbani Waters," in *Israel and Arab Waters: An International Symposium* (London: The Arab Research Center, 1984), 53.

10. A. H. Miller, "The Jordan River—Too Little for Too Many," Global Seminar, online, http://www.seagrant.wisc.edu/Advisory/Water_Seminar/Jordan_River/jordan_r.htm#Watershed (as of June 1999).

11. Thomas Naff and R. S. Matson, *Water in the Middle East* (Boulder, Colo.: Westview Press, 1984).

12. Thomas Stauffer, *Water and War in the Middle East: The Hydraulic Parameters of Conflict* (Washington, D.C.: The Center for Policy Analysis on Palestine, 1996), 9.

13. World Resources Institute, *Annual Global Environmental Data on CD*, 1992.

14. World Resources Institute, 1992.

15. Population Reference Bureau, *1996 World Population Data Sheet* (Washington, D.C.: Population Reference Bureau, 1996).

16. Gary Hoch, "The Politics of Water in the Middle East," *Middle East Insight* 9, no. 3 (March-April 1993): 20.

17. Abdul-Karim Sadik and Shawki Barghouti, "The Water Problems of the Arab World: Management of Scarce Resources," in *Water in the Arab World—Perspectives and Prognosis* (Cambridge, Mass.: Harvard University Press, 1994), 2.

18. World Bank, *Water Resources Management: A Policy Paper* (Washington, D.C.: The World Bank, 1993); World Bank, *World Development Report 1992: Development and the Environment* (New York: Oxford University Press, 1992).

19. M. Falkinmark and C. Widstrand, *Population and Water Resources: A Delicate Balance* (Washington, D.C.: Population Reference Bureau, 1992).

20. World Resources Institute, *World Resources 1996–97 Database* (Washington, D.C.).

21. Ines Dombrowsky, "The Jordan River Basin: Prospects for Cooperation within the Middle East Peace Process?" in *Water in the Middle East: Potential for Conflicts and Prospects for Cooperation* (Berlin: Springer, 1998), 96.

22. Sadik and Barghouti, "Water Problems," 5; "USAID: Country Health Statistical Profile for Jordan, 2001"; See also *U.S. State Department Factbook on Jordan*, available online at http://www.state.gov/r/pa/ei/bgn/3464.htm#people.

23. Drake, "Water Resource Conflicts," 5.

24. Peter Rogers, "The Agenda for the Next Thirty Years," in *Water in the Arab World—Perspectives and Prognosis* (Cambridge, Mass.: Harvard University Press, 1994), 293.

25. Central Bureau of Statistics, Government of Israel, Statistical Abstract of Israel, http://geography.miningco.com/gi/dynamic/offsite.htm?site=http://www.cbs.gov.il/engindex.htm, 1998.

26. Kally, *Water and Peace*, 52.

27. Hilary Andersson, "Dead Sea in Danger," BBC Online Network, August 29, 1999, http://news2.thls.bbc.co.uk/hi/english/world/middle%5Feast/newssid%5F392000/39242.stm.

28. Stauffer, *Water and War*, 4.

29. Stauffer, *Water and War*, 7.

30. Danna Harman, "Jordan: Israeli Decision to Cut Water Supply Casts Suspicion on Peace Process," *Jerusalem Post*, March 18, 1999.

31. Hassan Bin Talal, "Introduction," *Israel and Arab Waters: An International Symposium* (London: The Arab Research Center, 1984), iii–iv.

32. Paul Adams, "West Bank Water Row," BBC Online Network, August 26, 1999, http://news2.thls.bbc.co.uk/hi/english/world/middle%5Feast/newssid%5F58000/158955.stm (as of October 1999).

33. Nurit Kliot, *Water Resources and Conflict in the Middle East* (London: Routledge, 1994).

34. Drake, "Water Resource Conflict," 11.

35. Drake, "Water Resource Conflict," 11; Adams, "West Bank."

36. Hoch, "Politics of Water," 21.

37. Dombrowksy, "Jordan River," 96.

38. Lester R. Brown and Brian Halweil, "China's Water Shortage Could Shake World Food Security," *Worldwatch Institute* 11, no. 4 (1998): 10–11.

39. Jorg Barandat and Aytul Kaplan, "International Water Law: Regulations for Cooperation and Discussion of the International Water Convention," in *Water in the Middle East: Potential for Conflicts and Prospects for Cooperation* (Berlin: Springer, 1998), 14.

40. F. J. Berber, *Die Rechtsquellen des Internationalen Wassernutzungsrechts* (Munich, Germany: 1955), 13.

41. Berber, *Die Rechtsquellen*, 13.

42. Stauffer, *Water and War*, 9.

43. Leslie Schimda, "Israeli Water Projects and Their Repercussions on the Arab-Israeli Conflict," in Abdel Majid Farid and Hussein Sirriyeh, eds. *Israel and Arab Water* (New York: Ithaca, 1984), 28.

44. Priit J. Vesilind, "Middle East Water—Critical Resource," *National Geographic* 138, no. 5 (May 1993): 48.

45. Vesilind, "Middle East Water," 60–62.

46. Stauffer, *Water and War*, 8.

47. Stauffer, *Water and War*, 12.

48. Thomas Naff and Frederick Frey, "Water: An Emerging Issue in the Middle East?" *Annals, American Association of Political Science*, 431, Special issue, November 1985.

49. Frederick Frey, "The Political Context of Conflict and Cooperation over International River Basins," Conference on Middle East Water Crisis, Waterloo, Ontario, Canada, May 7–9, 1992.

50. A *National Security Strategy for a New Century* (Washington, D.C.: The White House, December 1999), 3.

CONCLUSION

The Environment as a Global Stability-Security Issue

Max G. Manwaring

When what mattered most in U.S. national security policy was military bases, preserving access to sea lines of communication, choke points, and raw materials—and denying those assets to the Soviet Union and its surrogates—the United States could generally ignore internal conditions in other countries. But, since the United States is now also interested in the need for nonhostile dispositions toward the country, the nonproliferation of weapons of mass destruction (WMD), the capacity of other countries to buy American-made products, the continued development of democratic and free market institutions, and human rights—as well as cooperation on shared problems such as illegal drugs, the environment, and the victims of natural and man-made disasters—then the United States and its allies must concern themselves with the internal conditions that spawn subnational, national, regional, and global instability.[1]

In these terms, the United States faces a challenge to change perspectives. The country needs an organizing paradigm to assist in clarifying its global leadership role, purposes, and courses of action. One message is unmistakable. The emerging global order has given the United States the longest period of economic prosperity anyone in the current generation can remember, but the end of the Cold War era conflict did not signal the end of all global conflict. Indeed, just the reverse is proving to be true. Moreover, the revolutionary nature of the global security environment has led to a more fragmented, complex, and explosive world than ever imagined. If the United States wants to preserve the present global prosperity and continue to benefit from it, the country must pay for it and nurture it.

Thus, U.S. interests, and those of the fragile and interdependent global community, demand a rethinking of contemporary security and stability. In general, this should be done with the clear understanding that non-military socio-economic, political, and environmental issues are integral parts of a new security-stability paradigm. In particular, this should be done with the clear understanding that environmental degradation is a major cause of instability that has implications for intra-national and cross-border conflict. Anything else would be irrelevant. In these terms, the United States must not think so much in terms of military 'war fighting' and traditional peace enforcement. Likewise, U.S. leadership should not dwell too much in the narrow confines of economic security, resource security, energy security, bio-security, or environmental security. Rather, the United States must think and act in the broad strategic context of global stability and security—as a war preventer and global stabilizer.

To help civilian and military leaders, and interested citizens, come to grips analytically with the implications of the realities of the contemporary global security environment, this chapter seeks to do several things. First, we put the environmental security issue into the context of the larger global stability-security issue. Second, we outline the violent characteristics of the new global security arena. Third, we analyze the complex environmentally stressed threat situation. Finally, we argue that substantially more sophisticated civil-military security-stability concepts and policy structures will be required for the United States to play more effectively in the contemporary security arena.

Environmental Security in the Context of Global Stability and Security

Before World War II, especially in the English-speaking countries, security had been almost exclusively the province of soldiers. Security was a term primarily associated with possible or probable military threats from other nation-states concerning strategic access or denial of raw materials, markets, lanes of communication, choke points, or the national territory. As a corollary, strategy was the use of military means to achieve the objects (i.e., ends) of national policy.[2]

After World War II and during the Cold War, a broader definition of security began to emerge. It involved the multi-dimensional political-economic-psychological—as well as military—confrontations between communist and non-communist powers.[3] And, as a corollary, the U.S. Department of Defense differentiated "military strategy" from "national strategy" by defining the broader term as "The art and science of developing and using the political, economic, and psychological powers of a nation, together with its armed forces, during peace and war, to secure national objectives."[4]

The current international security dialogue goes beyond traditional national policy objectives and focuses on relative well-being. Now, more and more, national security implies protection—through a variety of military and non-military means—of more ambiguous political, economic, social, and ideological interests.[5]

Additionally, the contemporary security dialogue stresses that challenges to the national well-being are generated by a lack of development and resultant chronic poverty, violence, and instability.[6] As a consequence, security can no longer be considered in terms of protecting national territory and interests against outside military aggressors. Rather, security is being defined as stability—and stability is dependent on the legitimate political, economic, and social development of the entire global community.

The primary implication of this broadened concept of security is that it is ultimately dependent on eradicating the root causes of instability. In these terms, ecological degradation is not simply a humanitarian concern. It is also the root cause of various types and levels of conflict, and poses an indirect and potentially serious threat to national well-being and international stability.[7] Environmental problems or stressors, thus, represent a nontraditional and complex national security threat that requires, first, a new look at the new global security arena.

The New Global Security Environment

If the appropriate magic could be conjured and one could look down through the familiar artificial political lines and colors of a current world map into the twenty-first-century strategic reality, one could see a complex new security environment. A deeper look into that picture would provide snapshots that show several types of ambiguous and uncomfortable conflicts, and state failure—and their causes and consequences. A few examples would include:

- A vision of twenty-six ongoing high-intensity wars, seventy-eight low-intensity conflicts, and 178 small-scale internal wars overlapping with the others. This picture would also show unspeakable human destruction and misery, and related refugee flows. In this connection, the magic snapshot would show that during the period since the Persian Gulf War, anywhere from 80 to 210 million people have lost their hopes, their property, and their lives.[8] The resultant political alienation—sufficiently reinforced by significant environmental degradation, economic and social deprivation, and spontaneous social violence—tends to direct the survivors and their advocates toward violence and the tactics of despair.
- A view of a vicious downward spiral that manifests itself in diminished levels of popular and institutional acceptance and support for weak and ineffectual governments and generates further disorder, violent internal conflicts, and mushrooming demands by various groups for political autonomy. These governance issues further translate themselves into constant subtle and not so subtle struggles for power that dominate life throughout much of the world today. Results of these dynamics can been seen not so much in the proliferation of a host of new countries, but in an

ironic explosion of weak, incompetent, misguided, insensitive, and/or corrupt governments throughout the world.

- In that connection, one can see a broken pattern of emerging city-states, shanty-states, amorphous warlord-controlled regions, criminal anarchist controlled regions, and a "steady run of uncivil wars sundering fragile but functioning nation-states and gnawing at the well-being of stabile nations."[9] This new vision of asymmetric battlefields and ambiguous internal wars reveals the human suffering created by weak and insensitive governance that spawns disease, poverty, crime, violence, and regional, national, and global instability. Ultimately, this instability—along with the destabilizers noted above—leads to a crisis of governance and another downward spiral into failing and failed state status.[10]
- Lastly, with a final adjustment of focus into the context of contemporary conflict and state failure, one can see a psychological state of mind in some individuals who think in terms of pure contempt for those outside their own small elite and ideological brotherhoods.[11] Evidence over time and throughout the world demonstrates that these types of self-appointed individual men and women are prepared to kill and destroy—and perhaps die in the process—to achieve their self-determined objectives.

The primary implication of the complex and ambiguous situations described above is straightforward. The contemporary chaotic global strategic environment reflects a general lack of legitimate governance in many parts of the world today. Instability thrives under those conditions. Instability and violence are the general consequences of unreformed political, social, economic, and security institutions and concomitant misguided, insensitive, incompetent, and/or corrupt (i.e., illegitimate) governance.[12] Thus, governance is the root cause and the central strategic problem in the current unstable security arena.

Corollary One. The results of real versus perceived moral incorrectness or malfeasance can be seen in governments unwilling or unable to provide basic services, to maintain decent roads, education, health, and other public services for all segments of its population. It can been seen in the inability or unwillingness of governments to provide basic personal and collective security and functional legal systems that protect civil and human rights, and a sense of societal equity. Illegitimate governance can be seen in governments unwilling or unable to deal fairly, effectively, or at all with environmental problems resulting from increasing competition for scarce resources such as arable land, forests, fisheries and other food sources, and uncontaminated water and air. It can be seen when disparate ethnic, religious, or other groups in a society insist on establishing separate identities, and government reacts with thuggish and brutal violence. Moral incorrectness or malfeasance can also be seen in governments unwilling or unable to deal effectively, fairly, or at all with displaced persons from internal degradation and violence, refugees from external famine, disease, or conflict, or roving bands of criminal anarchists.[13]

Corollary Two. Governments, international organizations, transnational entities, and other symbols of global power that have not been responsive to the importance of the legitimate governance reality find themselves in a "crisis of governance." They face growing social violence, criminal anarchy, and eventual destruction. Moreover, these governments are more susceptible than others to becoming pawns in the new "Great Game" that is being played for power and wealth between and among substate, state, and transnational actors. At the same time, the United States and the West confront a succession of failing and failed states, destabilized by internecine war.[14] These governance issues, then, are both causes and consequences of instability.

Corollary Three. Those who argue that instability and conflict—and the employment of terrorism as a tactic or strategy in conflict—is the result of poverty, disease, injustice, corruption, overpopulation, and misery may well be right. Evidence demonstrates, however, that those problems tend to be used to divert attention away from local governance issues to somebody or something else.[15] In any case, it is naïve to think that instability and conflict will disappear until the deeper human and political realities that produce poverty and misery are also confronted. It must be remembered that it is individual men and women—government leaders, civil and military bureaucrats, and transnational corporate leaders—who are ultimately responsible for confronting political, economic, and social injustice. And, it is individual men and women who react violently when a government or another symbol of power is perceived to be unable or unwilling to deal effectively with a given injustice. These individual men and women are also causes and consequences of instability.[16]

Conclusions. That takes us beyond the general instability threat to root causes—and to other consequences. That exposes the roots of global environmental degradation and reveals the foundations of injustice, violence, conflict, and state failure in the international system. That also provides a conceptual framework within which threat might be more completely understood, and more appropriate responses might be generated.

The Deeper, More Complex, and Multidimensional Threat Situation

Our analysis of the problem of contemporary conflict resolution and state failure indicates that if powers or coalitions of powers decide to become involved in an intranational stability problem, they should take the full responsibility of that action. That is, they normally should go beyond the "defensive" idea of "keeping a lid" on a problem in some conventional sense to the "offensive" idea of working toward a sustainable peace. One can easily envision situations in which U.S. interests might dictate acceptance of a less than ideal solution to a problem.[17] However, long-term interests as well as logic would normally dictate a durable peace rather than a crisis management solution.

Moreover, the general rule would be that decisionmakers and policymakers must carefully calculate the possible gains and losses, prioritize, and, when the case warrants, get involved earlier rather than later. If done earlier, this implies the initial and intense use of low-cost diplomatic and civilian resources and, perhaps, military support units. If done later, this normally implies the initial and intense use of high-cost military combat units to respond to a losing situation. Ultimately, however, the only viable test for involvement is clear national self-interest.

The first step in developing the strategic vision to conduct a responsible action to protect, maintain, or enhance some form of stability is to recognize the elements that define the *basic* threat and dictate response. The second step is to examine the question of appropriate short-, medium-, and long-term responses to a complex and multidimensional threat.

Basic Elements That Define Threat and Dictate Response. For traditionalists, the notion that a socionatural phenomenon such as environmental degradation could constitute a threat to national and global security is hard to accept. It has traditionally been perceived that ecological problems do not present either a direct or indirect threat to the people or territory of a nation-state. Generally, land squabbles, water rights controversies, looting, riots, conflict-driven migrations, and other violent consequences of environmental stresses are ascribed to religious, ethnic, and/or cultural conflict. But, it turns out that the roots of such conflicts—ostensibly political and socioeconomic—are interwoven with basic environmental problems.[18]

These consequences or symptoms of instability operate on several different levels. They range from natural, or man-made, or government-made disasters that create resource shortages and population stresses; to other actions that generate structural degradation and greater economic scarcity; to constrained economic production, disease, migration, and expulsion; to one level or another of violence that ultimately tends to result in major internal and external security problems.[19] This is best shown by a simplified causal linkage model (table 1) of environmental stress, security problems, violence, and eventual state failure.

In sum, table C.1 shows the links between population stresses and serious internal and international friction. It demonstrates the virtual impossibility of resolving instability problems without simultaneously dealing with the environmental causes that produce them in the first place. It also demonstrates a requirement for a new conceptual scheme, or paradigm, that takes into cognizance the complete strategic, or macrolevel, socioenvironmental multicausal picture in a coordinated and rational manner. Lastly, it indicates a serious need to adopt a holistic multidimensional multiagency approach for prevention, mitigation, or peaceful resolution of environmental stressors and environmentally derived conflicts.

Appropriate Short-, Mid-, and Long-Term Responses. The indirect and ambiguous challenges of ecological degradation are not easily countered by conventional security methods. As a matter of fact, in their early stages of development, a conventional security approach would be at best irrelevant, and at worst counterproductive. Appropriately countering the general instability threat—at its various stages of development—requires a nontraditional strategic vision and

Table C.1. Casual Links between Environmental Degradation, Security Problems, Violence, and Eventual State Failure

Stage One: The Beginnings of the Threat		
Resource Shortages Overexploitation of land, forest, and water resources. Production of industrial contaminants of land, forest and resources. Results in land, forest, water, and industrial slowing, etc. Generating rising wealth differentials.	**Resource and Environmental Stress, and Governmental Stresses.** Results in rising wealth differentials, societal and governmental officials taking advantage of shortages for own advantages, and boosting capital demand faced by the state and the economy. In turn increases susceptibility of government, economy, and society to sudden shocks like floods, droughts, and sharp downward changes in the economy, political system, and security environment.	Dealing effectively with these problems demands strong, competent, resilient, and uncorrupted governmental action at all levels of economy and society, e.g., legitimate goverance. Also, requires strong international coordination and cooperation. Otherwise, the environmental situation can move into Stage Two.
Population Stresses Reduction of standard of living and increased social violence. Results in internal and external migration, and social segmentation. In turn generates capital shortfalls and susceptibilty to sharp changes in the economy, the political system, and the security environment.		

(continued)

Table C.1. (*continued*)

Stage Two: Moving Closer to the Threat

Phase I–Precursors Threat	Phase II–Third-Level Threat	Phase III–Second-Level Threat
Resource and enviromental stresses, and governmental stresses incompetently or insensitively managed. Resulting in structural degradation and economic scarcity.	Constrained economic production, disease, migration, expulsion, elite rent-seeking (greed), and power-grabbing. Results in personal violence strikes, kidnappings, bank robberies, bombings, criminal anarchy and death squads, and refugee flows. Requires short-term political, social, and economic reforms; and minimum force to help ensure personal security.	Inadequate reforms and adherence to rule of law results in further social, economic, and political degradation, and more widespread and better social violence. Requires medium- to long-term deep reforms to deal with poverty and other root causes and the prudent application of force to enforce reforms and law and order.

Stage Three: Violent Conflict

Phase IV–First-Level Threat	Phase V–Failing or Failed State Syndrome
Regime unwillingness or inability to promulgate and enforce necessary reforms, and provide personal and national security results in further weakening of institutions, group identity crises and ethnic conflicts, more internal and external migration, terrorism, coups d'etat, insurgency, and possible external intervention. Requires continuation of reforms and development of deep regime legitimacy; as well as development of superior organization, unity of effort, and ability to exert deadly force against violent internal foe(s).	Failure to achieve legitimacy and deal effectively with progressively worsening internal social, economic, and security problems results in virtually complete turmoil and generally ineffective institutions. Thus, intervention to preclude state failure, or state failure.

approach. Moreover, the most well-intentioned reconstruction programs tend to amount to little more than triage in the absence of a more systematic approach to environmental degradation and its aftermath.

In these terms, it is helpful to think of the consequences of instability as third-level threats to national and international security. Root causes must be recognized as second-level threats. The inability or unwillingness of government to develop long-term, holistic, and morally acceptable means to maintain internal stability and development must be understood as the first-level (i.e., most fundamental) threat. As a result, strategic planners and decision makers must contemplate all three levels of threat in dealing with contemporary unconventional security-stability matters such as ecological degradation.

At the same time, another threat emerges at a fourth level that is both a cause and an effect of instability and violence. That is, once a violent internal adversary becomes firmly established, first-level reform and development efforts aimed at second-level root causes would be insufficient to control or neutralize a third-level (e.g., terrorist) threat. That third-level violent internal foe—regardless of whether or not it is sincerely trying to achieve specific political-economic-social-moral reforms, or only trying to gain some visceral satisfaction—can only be finally defeated by a superior organization, a holistic and unified strategy designed to promulgate deeper and more fundamental reforms—and carefully applied deadly force.[20]

As indicated in table C.2, the sum of the parts of an effective response equals: (1) the recognition at the highest levels of a destabilizing problem; (2) a sure capability to coordinate political, economic, social, and security objectives on the first and second levels; but also (3) to exert effective force at the third and fourth levels.

Table C.2. Threat Dimensions

Level of Threat	*Type of Threat*	*Civil-Military Actions to Confront These Threats*
3d (Third)	Instability; e.g., personal violence increased strikes, kidnappings, bank robberies, violent takeovers, death squads, bombings, murders/assassinations, criminal anarchy; and the beginnings of insurgency, ethnic cleansing, and refugee flows.	Short-term political, economic, and social reform; and minimum force to guarantee personal security.
2d (Second)	Widespread social violence; e.g., increases in third-level instability.	Medium- to long-term deep reforms to deal with poverty and other root causes of instability; and the careful application of force to enforce reforms and law and order.
1st (First)	Regime unwillingness or inability to promulgate and enforce second- and third-level reforms and provide personal and national security.	Long-term deeper reforms to create changes in mind-set; and development of political, military/police competence under an umbrella of legitimacy.
4th (Fourth)	Completely entrenched violent "revolutionaries."	Continuation of reforms and development of deep regime legitimacy; plus development of superior organization, unity of effort, and ability to exert deadly force against a violent internal foe.

Thus, civilian and military leadership must consider the logic of defining response in terms of the explicit and implicit "defensive" and "offensive" actions indicated by the model outlined in table C.1 and refined in table C.2. Defense action is relatively short term. It is primarily military-police and focuses on stopping parties to a conflict from shooting or moving against each other. Limited defensive and offensive (i.e., remedial) action can also be taken against causes as well as consequences of instability and violence. In environmental stability-security preconflict stages, appropriate responses might well be watershed-by-watershed, forest-by-forest, reef-by-reef, and shantytown-by-shantytown, but these responses will not be piecemeal and ad hoc. They will be a part of a coherent and prioritized whole effort.

Offensive action is generally medium to long term. It is primarily civilian and political-economic, but—from time to time—is likely to have to be coordinated with security organizations. It focuses on following up "defensive" action by attempting to diminish and remove the ecological, social, economic, and political causes of the instability and resultant violence. Additionally, this kind of "offensive" civil action can by applied early at the beginning signs of a threat without waiting for any kind of traditional "defensive" measures to be taken. These "offensive" actions can also initiate the steps necessary to prevent, mitigate, and resolve ecological stressors; and develop institutions, procedures, and attitudes necessary to the development of a legitimate civil society—and a durable peace.

The problems of environmental degradation and their concomitant consequences—such as poverty, hunger, disease, social violence, refugee flows, and conflict over scarce resources—easily flow across international frontiers. It is impossible to think in terms of a conventional unilateral approach being successful in countering any of these transnational instabilities. An appropriate effort to counter the causes and consequences of contemporary instability must include a relevant multinational component for effective coordination and cooperation.

Conclusions. Admittedly, putting the already large environmental security issue into an even larger global stability-security context generates such a complex problem that many leaders and scholars would be tempted to disregard this approach as "too big," "too hard," or "impossible." But, looking at the grand strategic picture allows one to better understand: (1) "that environmental issues can no longer be thought of as ancillary"[21] or dealt with in a piecemeal or ad hoc fashion; (2) that "substantially more sophisticated policy structures will be required to define and manage the interests of nation-states";[22] and (3) that what the United States needs is "an overarching campaign plan to operationalize strategic [shaping and engagement] guidance."[23]

Civil-Military Implications for Playing in the Global Security Arena

The United States is embroiled in a world of dangerous uncertainty in which time-honored concepts of national security and the classical military means to attain it, while still necessary, are no longer sufficient. Clearly, effective involvement in the contemporary global security environment requires conceptual and organizational adjustments. At a minimum there are at least three critical areas for interagency civil-military exploration and development. They are: (1) understanding and dealing with the political complexity and ambiguity of the stability-security equation; (2) addressing and resolving the problem of developing an "overarching" stability-security campaign plan; and (3) addressing and resolving the problem of "ad-hoc-ery" and the need for civil-military unity of effort.

Political Complexity and Ambiguity. Political complexity can been seen in several ways.[24]

- First, the ambiguous political-economic-psychological-moral nature of contemporary conflict situations forces the redefinition of long-used terms. The "enemy" is no longer a recognizable military entity or an industrial capability to make traditional war. The "enemy" is now the individual actor that plans and implements violence, and exploits the causes of violence for his own nefarious purposes. The "enemy" is now also poverty, disease, and other stressors that must be dealt with on their own terms. Thus, at base, the "enemy" is violence—and the causes of violence.
- Second, "power" is no longer combat firepower directed at a uniformed enemy military formation or industrial complex. "Power" is multilevel and combined political, psychological, moral, informational, economic, social, military, and police activity that can be brought to bear appropriately on the causes as well as the perpetrators of violence.
- Third, "victory" or success is not an "unconditional surrender" and signed formal document terminating a conflict. There is no easily identifiable human foe to attack and destroy, no specific territory to take and hold, no single credible government or political actor with which to deal, no guarantee that any agreement between or among contending authorities will be honored, and no specific rules to guide leadership in a given "engagement" process. Victory is now—more and more—and perhaps with a bit of spin control—defined as the achievement of a sustainable peace. Those who would declare victory and go home before achieving a sustainable peace need not apply.

- Fourth, conflict is no longer a military-to-military war of attrition. Conflict now involves entire populations. It now involves a large number of national civilian and military agencies, other national civilian organizations, international organizations, nongovernmental organizations, and subnational indigenous actors involved in dealing one way or another with myriad threats to security, peace, and well-being. The challenge of appropriate response is to recognize that the modern world is much too interrelated and complicated to allow a strictly economic or technical solution for mitigating and resolving national development problems, or to allow a strictly military solution to violence and instability in the shadowy gray area between peace and war.
- Finally, contemporary conflict situations are not limited; they are total. Contemporary conflict is not a kind of appendage—a lesser or limited thing—to the comfortable traditional vision of war. It is a great deal more. As long as opposition exists that is willing to risk everything to take down violently a government, destroy a society, or cause great harm to a society—there is war. As long as environmental stressors exist that can also lead to the destruction of a people, society, and/or government—there is conflict.

Ultimately, this is a zero-sum game in which there is only one winner. It is, thus, total.

In sum, these are the basic political realities for now and into the twenty-first century. These realities must inform the development of a conceptual framework—or filter mechanism—that can support issue identification, prioritization, and action to deal with the constellation of environmental and stability-security issues that supports or threatens the well-being of a people, society, or country.

Developing an Overall Architecture for a Global Stability-Security "Campaign Plan." Like George F. Kennan's Containment Theory of Engagement that governed the Cold War effort to deal with the hegemony of an expansionist Soviet Union, this blueprint for thinking and action would be intended to help decision makers, policy makers, and the public to understand the nature of the current global security environment. It does two additional things. First, it provides a conceptual scheme by which to cope successfully with the central strategic problem of governance. Second, the adroit application of this paradigm can create the quintessence of American pragmatism—a marriage of Wilsonian idealism and realpolitik. Thus, one may take a proactive approach to achieve something "better," rather than react defensively against something "bad."[25]

The analytical model proposed above in tables C.1 and C.2 provides decision makers and planning staffs a "filter mechanism" to guide thinking, planning, and implementation by focusing on the principal concepts that one must consider

their staffs must analyze the problem as a whole and develop the vision necessary for complete success.

The Main Challenge, Threat, and Task

Over the past several years, it seems that many decision makers, policy makers, and opinion makers have been consistently surprised at the chaos and violence that stems from environmental degradation and concomitant stressors. They have also been confused and unable to decide what to do or how to do it beyond the usual "crisis management" and "spin control" tactics. This kind of piecemeal and ad hoc approach to contemporary instability problems brought on by environmental stressors reminds one of the following statement:

> The trouble in Ivory Coast could be seen coming for at least five years, visible on the horizon like the darkness of a gathering storm. But, as predictable as the disputed vote and killings may seem, they were never inevitable.[27]

The Challenge. The challenge, then, is to come to terms with the fact that contemporary security and stability—at whatever level—is at base a "governance" issue. The corollary is to change from a singular military to a multidimensional political-economic-psychological-security paradigm.

The Threat. The ultimate threat is that unless and until leaders at the highest levels recognize what is happening strategically, reorient thinking and actions appropriately, and are able to educate and lead their constituencies into the reality of the post-Cold War world, the problem of global stability will resolve itself. There won't be any.

The Main Task. The main task is to construct stability and a sustainable peace on the foundation of a carefully thought-out, holistic, long-term, phased planning and implementation process that focuses on the five elements that make up the stages that depict the causal links between environmental degradation, security problems, violence, and eventual state failure outlined in tables 1 and 2, above. They take into account the major unifying and legitimizing dimensions of contemporary stability-security requirements. Hopefully, they will reinitiate a process of rethinking both problem and response.

Afterword

Implementing the extraordinary challenges of reform and regeneration implied in this call for a paradigm change will not be easy. That will, however, be far less demanding and costly in political, military, and monetary terms than continuing a singular crisis management and generally military approach to global stability and security that is inherently a long-term political problem.

The special status of the United States allows it the opportunity to facilitate positive change. By accepting this leadership challenge, this country can help replace conflict with cooperation and harvest the hope and fulfill the promise that global stability offers.

Notes

1. *A National Security Strategy of Engagement and Enlargement* (Washington, D.C.: The White House, February 1996); *A National Security Strategy for a New Century* (Washington, D.C.: The White House, May 1997); *A National Security Strategy for a New Century* (Washington, D.C.: The White House, October 1998), etc., to date. Hereafter cited as White House.

2. Russell F. Weigley, "The Evolution of Strategic Thought," in B. Thomas Trent and James E. Harf, eds., *National Security Affairs: Theoretical Perspectives and Contemporary Issues* (New Brunswick, N.J.: Transaction Books, 1982), 69–71.

3. Weigley, "Evolution."

4. DOD, *Dictionary of U.S. Military Terms for Joint Usage* (Washington, D.C.: U.S. Government Printing Office, 1974).

5. Amos A. Jordan, William J. Taylor, Jr., and Lawrence J. Korb, *American National Security: Policy and Process*, 3d ed. (Baltimore, Md.: Johns Hopkins University Press, 1988), 3.

6. White House.

7. White House.

8. The data in this note can be seen in "World Conflict and Human Rights Map 2000," prepared by PIOOM for IIMCR with the support of the Goals for Americans Foundation, St. Louis, Mo.

9. Leslie H. Gelb, "Quelling the Teacup Wars," *Foreign Affairs* (November-December 1994): 5.

10. Daniel C. Esty et al., "The State Failure Projects: Early Warning Research for U.S. Foreign Policy Planning," in John L. Davies and Ted Robert Gurr, eds., *Preventive Measures: Building Risk Assessment and Crisis Early Warning Systems* (New York: Rowman & Littlefield, 1998), 27–38. Also see Thomas F. Homer-Dixon, "On the Threshold: Environmental Changes as Causes of Acute Conflict," *International Security* (Fall 1991): 76–116; and Thomas F. Homer-Dixon, *Environment, Scarcity, and Violence*, (Princeton, N.J.: Princeton University Press, 1999), 1238–168.

11. Robert D. Kaplan, "The Coming Anarchy," *Atlantic Monthly*, February 1994, 72–76; and Robert D. Kaplan, *The Coming Anarchy* (New York: Random House, 2000), 3–57.

12. Esty et al., "State Failure," Davies and Gurr, "Preventive," and Homer-Dixon, "Threshold."

13. *Ibid.*

14. *Ibid.*

15. *Ibid.*; and Kaplan.

16. Martha Crenshaw, "The Causes of Terrorism," *Comparative Politics* (July 1981): 379–99; and Martha Crenshaw, ed., *Terrorism in Context* (University Park, Pa.: Pennsylvania State University Press, 1995).

17. For an elaboration of the "offensive" and "defensive" ideas, see David Last, "Winning the Savage Wars of Peace: What the Manwaring Paradigm Tells Us," in John T. Fishel, ed., *The Savage Wars of Peace* (Boulder, Colo.: Westview Press, 1998), 211–39.

18. Esty et al., "State Failure" and Homer-Dixon, "Threshold."

19. *Ibid.*

20. General Sir Robert Thompson was one of the first to note this phenomenon as a result of his experience in Malaysia. Author interview, January 16, 1986, in Washington, D.C. Also see: Robert Thompson, *Defeating Community Insurgency* (New York: Praeger, 1966), 87.

21. Braden R. Allenby, "Environmental Security: Concept and Implementation," *International Political Science Review* (2000): 7.

22. Allenby "Environmental Security," 6.

23. Author interview with General Charles E. Wilhelm, USMC (Ret.), on February 9, 2001, in Reston, Va.

24. This set of "realities" was first published in a different format by the author as "Thinking Again about Contemporary Conflict," in *The Officer*, March 2001, 34–36.

25. George F. Kennan, "The Sources of Soviet Conduct," *Foreign Affairs* (July 1947): 566–82.

26. Wilhelm, interview. Also: author interviews with General Anthony L. Zinni, USMC (Ret.), on June 2, 1999, and October 6, 2000, in Washington, D.C.

27. Howard W. French, *New York Times*, October 29, 2000, sec. 4, p. 1.

Index

About the Contributors

Dr. Stephen Blank is a research professor of National Security at the U.S. Army War College Strategic Studies Institute. He has his Ph.D. from the University of Chicago and writes on security issues pertaining to the Soviet bloc and the post-Soviet world since 1989. He is the author or editor of numerous articles, monographs, and books dealing with security issues. Dr. Blank's most recent article specifically addressing environmental security concerns is "A Sacred Place Is Never Empty: The External Geopolitics of the Trans-Caspian," in *Proceedings from the Conference on Natural Resources and National Security: Sources of Conflict and the National Interest*, JINSA, 2001.

Mrs. Darci Glass-Royal is CEO of GRS Solutions, Inc., a company based in northern Virginia that specializes in global economic and security issues. She is a graduate of Bryn Mawr College and George Washington University. She has conducted numerous, in-depth analyses of environmental problems in Central America, the Caribbean, South America, Africa, and Central Asia. In that connection, Mrs. Glass-Royal has worked directly for the Department of Defense, the State Department, and the U.S. Agency for International Development.

Captain Stephen D. Kiser, USAF, is a political science instructor at the U.S. Air Force Academy and is attached to the RAND Corporation in Santa Monica, California. He holds an M.A. degree from the University of Hawaii where he focused on future security issues in Asia. His academic interest lies in international security issues, with an emphasis on the emerging field of environmental security. Captain Kiser is the author of *Water: The Hydraulic*

Parameter of Conflict in the Middle East and North Africa, U.S. Air Force Academy Institute for National Security Studies, 2000.

Dr. Max G. Manwaring is a research professor of Military Strategy at the U.S. Army War College Strategic Studies Institute and a retired U.S. Army colonel. He holds a Ph.D. from the University of Illinois at Champaign-Urbana and writes on the theory of grand strategy, security policy, and political-military affairs in Latin America. He is the author, editor, or coeditor of several articles, monographs, and books dealing with strategy and security. Manwaring's most recent effort in the realm of security policy and strategy is *The Search for United States Security: A Forward Strategy for the Twenty-First Century*, Praeger, 2002.

Ambassador Frank McNeil, Department of State (Ret.), is a former United States ambassador and senior State Department official who has worked for years in Latin America and Asia. He is also a former director of the Navy's Strategic Studies Group. Now, a senior research associate of the North-South Center at the University of Miami, Ambassador McNeil serves on the Chief of Naval Operations Executive Panel and as a founding trustee of CR-USA, a foundation that provides major support to the environment in Costa Rica. His numerous writings include *War and Peace in Central America* (on the *New York Times* 1989 list of "notable books") and a groundbreaking essay, *Making Sense of Environmental Security*, University of Miami North-South Center Press, 2000.

Commander Ray Simmons, U.S. Navy (Ret.), is a graduate of the U.S. Naval Academy and Florida State University and holds degrees in both physics and physical oceanography. He has served inter alia as the Department of Defense lead to the Gore-Chernomyrdin Commission Environmental Working Group, director of the U.S. National Ice Center, and executive assistant to the vice chairman of the Joint Chiefs of Staff. Commander Simmons is currently the chief operating officer of GRS Solutions, where he also serves as director of research and studies with a primary focus on environmental and security analyses. In that connection, he has conducted numerous, in-depth environmental security studies for several U.S. government agencies and has lectured widely on the subject both in the United States and abroad.

Dr. Aondover A. Tarhule is an assistant professor of geography at the University of Oklahoma. He is a native of Nigeria, holds a Ph.D. from McMaster University in Canada, and received a postdoctoral research fellowship from Queens University in Canada. He writes on hydro-climatic and water resources issues.

Dr. John P. Warren is a retired senior Foreign Service officer with twenty-five years of experience with the U.S. Agency for International Development (USAID). He holds a Ph.D. from Texas A&M University and specializes in natural resources management, including sustainable agriculture, forest and watershed resources, and fisheries. In that connection, Dr. Warren has lived and worked in Africa, the Middle East, Central and South America, and the Caribbean. He has published over sixty articles and monographs on development and natural resources economics and policy. The most recent is a "Dominican Republic Environmental Assessment," International Resources Group, Ltd., for USAID/Santo Domingo, 2001.

General Charles E. Wilhelm, USMC (Ret.), is a former deputy assistant secretary of defense, commander of Marine forces in the Atlantic, Europe, and South America, and commander in chief of the United States Southern Command. While at the United States Southern Command, General Wilhelm directed U.S. military operations in the Caribbean and Central and South America, including U.S. relief efforts in response to Hurricane Mitch. He is an advocate of early and strong preventive and mitigative measures as they pertain to man-made and natural disasters.

General Anthony L. Zinni, USMC (Ret.), is a former commander in chief of the United States Central Command. While at the Central Command, General Zinni directed U.S. military operations in parts of Africa and Southeast Asia. He is a strong advocate of early diplomatic, preventive, and mitigative measures as they pertain to ecological degradation and stressors. At present General Zinni is the special United States envoy to the Middle East peace process.

www.ingramcontent.com/pod-product-compliance
Lightning Source LLC
LaVergne TN
LVHW102006210926
848483LV00009B/80
* 9 7 8 0 7 3 9 1 0 4 4 7 7 *